CW01471798

Weathering the Reformation

Weathering the Reformation explores the role of the Little Ice Age in early modern Christian culture and considers climate as a contributing factor in the Protestant Reform. This book focuses on religious narratives from Strasbourg between 1509 and 1541, pivotal years during which the European cultural concept of nature splintered along confessional differences. Together with case studies from antagonistic religious communities, Linnéa Rowlatt draws on annual weather reports to explore a period during which the climate became less hospitable to human endeavours. Social unrest and the cultural upheaval of Reform are examined in relation to deteriorating climatic conditions characteristic of the Spörer Minimum. This book will be of particular interest to scholars of religious history and climate history.

Linnéa Rowlatt is Research Coordinator for the International Network for Training, Education, and Research on Culture (Network on Culture) based in Canada. Her PhD was jointly awarded by the University of Kent, UK and the Freie Universität Berlin, Germany.

Routledge Studies in Religion and Environment
Series editor: Evan Berry

The Routledge Studies in Religion and Environment book series explores religious encounters with environmental challenges and strives to capture the ecological dimensions of religious life with empirical and theoretical sophistication. Resisting the urge to concentrate exclusively on religious traditions, this series conceives the term 'religion' broadly, seeking to include not only religious actors, institutions and theological traditions, but also lived spiritualities, Indigenous cosmovisions, para-religious organizations, and socially enacted notions of the sacred. Environmental challenges are manifest in every part of the world, but the bearing of religious actors, ideas, and institutions on these challenges is variable. Accordingly, this series is ambivalent about whether and how religion matters with respect to environmental issues. We welcome scholarly contributions that chart the dynamic relationships between systems of human meaning-making and environmental processes at all scales, from the planetary to the parochial.

Church, Cosmovision and the Environment
Religion and Social Conflict in Contemporary Latin America
Edited by Evan Berry and Robert Albro

Global Religious Environmental Activism
Emerging Conflicts and Tensions in Earth Stewardship
Edited by Jens Koehrsen, Julia Blanc and Fabian Huber

Weathering the Reformation
Climate and Religion in Early Sixteenth-Century Strasbourg
Linnéa Rowlatt

For more information about this series, please visit: https://www.routledge.com/Routledge-Studies-in-Religion-and-Environment/book-series/RSRE

Weathering the Reformation

Climate and Religion in Early
Sixteenth-Century Strasbourg

Linnéa Rowlatt

Routledge
Taylor & Francis Group

LONDON AND NEW YORK

First published 2024
by Routledge
4 Park Square, Milton Park, Abingdon, Oxon OX14 4RN

and by Routledge
605 Third Avenue, New York, NY 10158

Routledge is an imprint of the Taylor & Francis Group, an informa business

© 2024 Linnéa Rowlatt

The right of Linnéa Rowlatt to be identified as author of this work has been asserted in accordance with sections 77 and 78 of the Copyright, Designs and Patents Act 1988.

All rights reserved. No part of this book may be reprinted or reproduced or utilised in any form or by any electronic, mechanical, or other means, now known or hereafter invented, including photocopying and recording, or in any information storage or retrieval system, without permission in writing from the publishers.

Trademark notice: Product or corporate names may be trademarks or registered trademarks, and are used only for identification and explanation without intent to infringe.

British Library Cataloguing-in-Publication Data
A catalogue record for this book is available from the British Library

ISBN: 9781032193915 (hbk)
ISBN: 9781032201399 (pbk)
ISBN: 9781003262411 (ebk)

DOI: 10.4324/9781003262411

Typeset in Sabon
by codeMantra

Contents

Illustrations

Figures

Maps

Preface

The idea that there might be a reciprocal relationship between climate and religion confronts a long-held separation of the material and spiritual realms that is precious to Western culture. After centuries where God's voice bore a startling resemblance to that of powerful men, there are good reasons for this separation. However, as much as this is a contentious perspective today, considering the natural environment as having a role in religious and cultural developments is an unavoidable conclusion because our human bodies are inescapably part of the natural realm and subject to its laws, as are the material infrastructures we build. As culture-making creatures, we collectively seek to align our understanding about the essential nature of reality with our material experience; while cultural norms may fail the odd individual without shifting, when the majority of a population does not find their experience aligning with cultural norms, change in the culture usually ensues. One such moment of change was the Protestant Reformation, which launched during the zenith of the Spörer Minimum.

This book is directed towards several audiences: students of environmental history, religion, anthropology, and environmental sciences, particularly climatology, and any who are interested in exploring the conceptual roots of our current environmental crisis. I bring research from historical climatology, which is the science of understanding past climates, into relationship with research from the history of religion and the social history of Strasbourg and Alsace. I anticipate that some information shared herein may appear rudimentary to various readers and difficult to comprehend for others, with reversed perspectives quick to arrive. Nevertheless, as impressed upon me early in my career, reality does not exist in tidy academic silos and a work of interdisciplinary research must bring diverse fields of knowledge into relationship with each other. Hopefully I have done so in a manner which clarifies, rather than obscures, the relationship.

Much like a baby, most of the work in conceiving and delivering a book occurs in private; the questions which led to this particular book started demanding my attention while I was operating a biodynamic farm in Ottawa's Greenbelt. This volume is among the results of my quest to understand the cultural roots for Western society's self-destructive approach to the natural environment. However, books cannot flourish without the assistance of a small village, and I would like to express my gratitude to those who helped me with this project. Intellectually, I am indebted to Professor Marie-Françoise Guédon for helping me understand the nature of culture and to Professor Emeritus Richard C. Hoffmann for methods which unveil the relationship between culture and nature. Their instruction years ago is the skeleton of my research.

The core of this book is based on my doctoral research, conducted under the roof of the Text and Event in Early Modern Europe (TEEME): An Erasmus Mundus Joint Doctorate programme. Joining this programme in 2011 gave me a scholarly family within which I could share the ebbs and flows of doctoral research; being selected made the entire project possible. By insisting that research on the past be placed in a meaningful relationship with present issues, TEEME's architects ensured that new methodologies could be brought to bear on familiar stories, thereby delivering a fresh understanding of their relative place in history. Professor Bernhard Klein (Kent) and Professor Sabine Schülting (FU Berlin) were efficient and gracious anchors as the local TEEME coordinators. Professor Klein's assistance in winning a work placement with Doctor Farah Karim-Cooper in the research department of Shakespeare's Globe Theatre was a particular treasure he bestowed upon me, for which I am very glad. My two supervisors, Kenneth Fincham (Kent) and Claudia Ulbrich (FU Berlin), gave encouragement, guidance, discipline, and criticism as appropriate, and I will always be grateful for Ken's kindness when the stresses of the final stretch were overwhelming. Thank you, as well, to the members of my examination committee, whose comments and suggestions sharpened my initial findings into a more coherent argument.

Translation of primary source material has been an interesting challenge, and I cannot claim to be an expert in all the languages that are included in this book. In particular, as someone whose skills in German remain rudimentary, I am deeply grateful to Rose Fuhrmann, Volker Honemann, and Jocelyn Egginton for their collaboration in translating *Alemmanisch Mittelhochdeutsch* (Alemannic Middle High German) passages. I am also surpassingly grateful to Susannah Brower for her gracious assistance with Latin translations in much of the book and to Andrew Reeves for heroically stepping in at the last moment with assistance for the Latin translations in Chapter 4. Any errors which may remain now, whether in translation or elsewhere, are my own responsibility.

Lisa Crandall has been a true ally in this venture. I am very grateful for her courageous generosity in reading through all the drafts and the insight of her comments. I am also grateful for the improvements offered by Jolanta Komornicka to my index. Another ally was John Bowra, who extended his neighbourliness to regular meals and good cheer. Kristine Nickel and Jeff Gray have had faith in my capacities from the beginning and my success is at least partly due to the strength of their vision. Thomas Lips, my husband, is unstinting in his support, encouragement, and patience; his editorial skills benefit not only myself but also anyone who opens this book. My daughters, Karmen and Freija Walther, have been cheering for me from the beginning and I am very grateful for their support.

Introduction

Questions about the fact of life are a concern of theology as well as natural science and, during the uncounted centuries when the two were considered a single pursuit, were unavoidably entwined. Out of a need to respond to such questions, religious explanations for the existence and structure of the natural world developed into wide-ranging discourses along with guidance for a virtuous human attitude and approach to it. Religious debate exerted and continues to exert influence over such considerations as the origin, structure, and nature of the universe, the nature of being, the development of suitable ritual practice to induce advantageous environmental conditions and to prevent or to address unexpected and unwanted extreme weather events, and the establishment of boundaries between orthodox and heretical ideas about nature.

The passage of time brings change to religious convictions; beliefs and opinions which once exerted great cultural influence are eclipsed, with new and different views drawing attention and resources and, whether quickly or slowly, eventually finding expression through social behaviour. The natural environment is also forever in motion, with processes, events, and forces which actively influence the material conditions for human life. The climate was changing in early modern Europe: waves and troughs of the Little Ice Age, such as the Spörer Minimum which frames this research, manifested as unreliable and uncertain weather conditions all over Europe. It was also a period when the traditional interpretative authority of the Roman Catholic Church was called into question and successfully challenged. Although the movement is referred to as the Protestant Reformation (since proponents initially aimed only to reform what they considered to be a corrupt institution), in fact at least two new interpretations of the Christian message were born during this time: the radical interpretation, later known as Anabaptism, and the Evangelical Reform interpretation, later known as Protestantism.

In essence, this book explores the relationship between the climate and Christian culture at a point in time and space when both nature and culture were in turmoil. Chapter 1 begins with a brief physical description of the Upper Rhine Valley and an outline of the social conditions and tensions in early

DOI: 10.4324/9781003262411-1

sixteenth-century Alsace. Chapters 2–4 provide a 67-year weather report broken into three sections (1473–1509, 1509–1525, 1525–1540) while also describing some of the social responses of Alsatians and Strasbourgeois to the weather conditions. The weather report serves as a thread of continuity for these chapters, as, along with each section of the weather report, I explore the dominant religious cultures in Alsace and Strasbourg during those years (Roman Catholic, radical, and Reform). Each exploration includes a case study foregrounding a particular Christian view of nature: Johannes Geiler von Kaysersberg's sermons in *Die Emeis* for an orthodox Roman Catholic perspective; *Ein Fast Schon Büchlin*, written by Strasbourg gardener Clemens Zyegler, for the radical perspective; and Wolfgang Capito's *Hexemeron Dei opus* for the Reform representation of the natural environment. The three case studies illuminate developments in a fragmenting Christian concept of the natural environment, since each one reveals a unique view about the essential nature of nature, how nature operates, and a confessionally appropriate approach to it.

On the basis of these explorations, I argue for two hypotheses in this book of climate history. First, that the undependable weather conditions characteristic of the Spörer Minimum combined with a rigid social hierarchy to exert such severe economic and social stress on people in Alsace and the surrounding regions that the weather contributed to their enthusiastic acceptance of Martin Luther's religious insights, leading to what we now know as the Protestant Reformation. Accordingly, climate must be considered a contributing factor to the success of the Protestant Reformation. Second, that the weather-related social turmoil of the period was the crucible for, among other theological developments, an authoritative new Christian representation of nature which erased inherent value from the natural world and represented God as having given it to humans for exploitation. Consequences arising from this sixteenth-century religious development can be seen in our contemporary Western culture, whose dominant worldview assumes entitlement to an endless exploitation of the natural environment.

Ultimately, drawing upon the hybrid and interactive model of socioeconomic metabolism (see below), I postulate that stress from the natural sphere of causation contributed to spontaneous developments in the Christian representation of that same natural sphere, and that these developments were integrated into the new Protestant message. The resulting revised representation of the natural environment has guided and continues to guide Protestant attitudes and approaches to the natural environment; as Protestantism accompanied European colonial expansion, this perspective has also shaped Western views of the environment and behaviour towards the natural world. The limitations of this perspective are clearly visible today and it behooves us to clarify the reasons it was adopted enthusiastically by early modern Evangelicals.

Context

This work is not the first research project to look into the relationship between religion and views of the natural world; historians of science, religion, and numerous other sub-fields have explored the issue in some depth. A prominent view is that offered by Max Weber, who, shortly after the First World War, began to speak and write of the secular perception of the natural environment as a particular privilege enjoyed by modern man. In a 1922 essay, he declared that thanks to increased intellectualization and rationalization,

> ... there are no more mysterious incalculable forces that come into play, but rather that one can, in principle, master all things by calculation. This means that the world is disenchanted. One need no longer have recourse to magical means in order to master or implore the spirits, as did the savage, for whom such mysterious powers existed. Technical means and calculations perform the service. This above all is what intellectualization means.[1]

Weber's thesis is founded on the view that such 'disenchantment' was the result of a religion that favoured a rational approach to the world: Protestant Christianity. This influential theory has been widely repeated and interpreted; that it is still in current usage is seen recently in, for example, Tom Scott's 2013 collection of essays *The Early Reformation in Germany: Between Secular Impact and Radical Vision*. Expanding on Robert Scribner's formulation of Weber's thesis as "the transformation of the sacramental to an anti-sacramental view of the material world,"[2] Scott suggests that such a mental turn is closely associated with anticlericalism, itself a product of Protestantism's "rejection of the visual, emblematic, and symbolic in favour of the ostensibly more rational written or spoken word."[3]

However, Weber's theory of the 'disenchantment of the world' has been deeply challenged since the last quarter of the twentieth century, as part of what Alexandra Walsham refers to as a 'bold revisionist backlash against the confident teleologies and polarities embedded in older models of modernization.'[4] Weber's argument about Protestantism is now understood to have been one step in his summary of a linear and triumphant progress to modernity, the particular heritage of Western civilization. Long promoted by Protestant theologians, this argument rested upon a representation of Roman Catholic sacraments as empty rituals promoting a magical view of the world which could, smugly, be dismissed as mistaken. Critics of the thesis, Scribner among them, re-frame the question as one of cultural continuity rather than an abrupt rupture based on the new religious movement; this view has set the tone for historical discourse on the issue. Scholars pursuing this analysis, however, largely seek to demonstrate that an 'enchanted' or 'magical' worldview was not eradicated by the new religious movement through demonstrating that Protestantism engendered rituals and a certain 'magic' of its own.[5]

While my book rests comfortably within this scholarly repudiation of Weber's 'disenchantment of nature' thesis, rather than seeking to confirm the continuity of a magical worldview into the Protestant era, my research challenges Weber on the basis that a weak form of rationality with respect to the natural world is found in the views of an orthodox theologian of the Roman Catholic community in Strasbourg (see Chapter 2). Moreover, the earliest Protestants were not consistent: while Wolfgang Capito declared that God gave the Earth to humans for our own purposes, both he and Jean Calvin wrote that God was completely, thoroughly imminent in and acting upon the natural environment (see Chapter 4). I am not alone in eroding the basis of Weber's comparison between enchanted and secular Christian views of nature; in particular, my argument builds upon Richard Kieckhefer's 1994 article 'The specific rationality of medieval magic,' Edward Grant's collection of essays *The Nature of Natural Philosophy in the Late Middle Ages,* and Stuart Clarke's magisterial 1997 monograph, *Thinking with Demons.*[6]

Lynn White, Jr., a medieval historian of technology, wrote a trenchant criticism in 1967 of Christian views of the natural environment entitled 'The Historical Roots of our Ecological Crisis.' White's brief article indicts Christianity as the intellectual and religious foundation of the current environmental crisis, accusing specifically Western Christianity of being the most anthropocentric religion the planet has ever known for insisting "that it is God's will that man exploit nature for his proper ends."[7] The article continues to provoke passionate debate, often between a defensive Christian community and secular critics. Since my research examines precisely the question of the Christian view of nature, albeit when the Christian community was in upheaval, it is unavoidable that my findings will contribute to the debate. In brief, my research confirms a theological argument of the early sixteenth century which conceptually represented Alsatian peasants as engaged in God's work when farming. Such a view is expressed in a pamphlet from Strasbourg gardener Clemens Zyegler published shortly before the 1525 German Peasants' War; a lengthier discussion of the material is available in Chapter 3. My research also confirms the emergence of an explicit Reform representation of the natural environment as having been created by God for human exploitation; this view was articulated by leading Strasbourg Reformer Wolfgang Capito in his exegesis of Genesis *Hexemeron Dei opus* (see Chapter 4).

Following an influential article published in 1999, Wolfgang Behringer published *Kulturgeschichte des Klimas* in 2007. In it, he outlines recent research in historical climatology from the onset of the Holocene until the medieval warm period, then focuses more tightly on the symptoms and effects of the Little Ice Age (LIA) (physical and social, then cultural). He finishes the book with an exploration of global warming, accepting the reality of the phenomenon but suggesting that alarms have been exaggerated. However,

Behringer largely limits his research into the cultural repercussions of the LIA to the last decades of the sixteenth century and the seventeenth century. While his research is comprehensive and fruitful, bringing forth interesting insights about, for example, the incidence and conditions of trials for witchcraft throughout Europe and melancholy as a mental illness typical of the LIA, most of Behringer's attention is on the years after 1560. My research takes steps into hitherto uncharted territory by scrutinizing the relationship of weather with social events and economic problems of the early sixteenth century, juxtaposing them with the cultural changes introduced with the Reformation, and exploring the new religious views for their new representations of the natural world.[8]

Several comprehensive studies of the role of climate in history have been published recently, including John L. Brooke's *Climate Change and the Course of Global History: A Rough Journey* (2014), Lieberman and Gordon's *Climate Change in Human History: Prehistory to the Present* (2018), White, Pfister, and Mauelshagen's edited volume *The Palgrave Handbook of Climate History* (2018), and Pfister and Wanner's *Climate and Society in Europe – the Last Thousand Years* (2021). Thematically closest to the present study, though, is Mike Hulme's *Weathered: Cultures of Climate* (2017). In it, Hulme provides a masterful and wide-ranging exposition of cultural meanings for the term *climate*, exploring how the concept mediates between our human experience of the passing weather and our cultural ways of living which are animated by this experience. He examines how diverse cultures understand, accommodate, blame, fear, and, among other attitudes and approaches, seek to predict and govern the climate. According to Hulme's categorizations, Alsatians' knowledge of climate can be considered an indigenous knowledge because their knowledge was mediated by a respected text (the Bible). As I show in this volume, when existing interpretations of that text (Roman Catholic interpretations) failed to provide protection from unreliable and unfavourable weather, the failure contributed to a choice for other interpretations of the text.[9]

Pace Hulme's research showing that *klima* was a concept available to ancient Hebrew and Classical Greek writers, I did not find the concept mentioned in the written discourse of Christian people in Strasbourg and Alsace in the early 1500s. However, some did jot down the weather they experienced, and theologians were definitely concerned with understanding the source and operations of the natural world, sharing their understandings with the faithful, and offering guidance about how human beings could please God in their approach to nature. To be true to my sources, then, my research gives an in-depth exploration of the concept of *nature* in three competing religious perspectives prevalent in Strasbourg and Alsace between 1509 and 1540.

The second, third, and fourth decades of the sixteenth century were the generative years for two new Christian interpretations of Scripture: Protestant and radical/Anabaptist. Theologians of both persuasions, as well as Catholic theologians, continued to develop distinct views of God's wishes for humanity in the centuries to follow. Examining the work of specific individuals from the three traditions during a particular time frame for significant differences in the representation of the natural environment is a small but worthwhile contribution towards understanding the conceptual and intellectual foundations of a worldview which disregards ecological balance as a structural priority. It should be noted that although the authors of the primary source material used herein may have been influential, there is no guarantee that their views were widely held, or even that these documents represent the authors' own sincerely held views over time. Acknowledging such uncertainties, the research offered here is based on a fundamental agreement with the assertion of Thomas Kirchhoff and Ludwig Trepl that there are significant differences in the ways we have historically considered and evaluated nature, and that an examination of changing views of nature as part of developing a comprehensive cultural history of humans on earth is valuable to our future.[10]

The hybrid and interactive model of socio-economic metabolism

Guiding this descriptive, analytical, and interpretive historical research project is the hybrid, interactive model of socio-economic metabolism. A relatively recent theoretical development in the history of thought about nature and humanity, this model recognizes the autonomous activity of both the natural world and humanity's cultural world and links them together at the site of human biophysical experience (see Figure I.1; this diagram serves

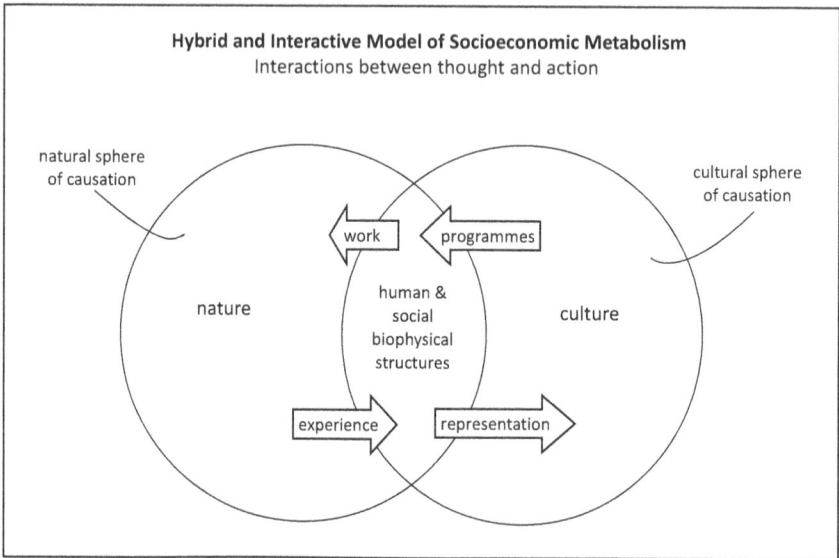

Figure I.1 Hybrid and interactive model of socio-economic metabolism demonstrating the flows of energy and materials between symbolic culture, material culture, and the natural environment.

only as heuristic simplification). Developed in the late twentieth century by Marina Fischer-Kowalski and Helga Weisz at the School of Social Ecology of the University of Vienna, the model is based on the recognition that human bodies, while qualitatively constructed through culture, are simultaneously biological organisms operating within the laws of nature and affected by the natural environment where they exist.[11]

In this theoretical framework, societies (including both human and social biophysical structures) are understood as the means of reproducing human populations, the primary material component of which are the bodies of the people. The total biological metabolism of any society will be the sum of a complex relationship between the body weight and reproductive rate of the humans, their working hours and the energy spent per working hour, as well as the material components of the natural world over which the culture assumes authority, such as domesticated animals, and those it takes responsibility for reproducing, such as crops. Smaller factors include the climate and other environmental circumstances, and human-made and maintained technical structures. As Richard Hoffmann points out, humans not only interact with nature, they consciously seek to use elements of it for their culturally determined purposes – to colonize it – and in so doing, modify the natural world whether intentionally or not.[12]

A long-term and slow co-evolutionary process ensues, engaging both the natural world (through natural evolution) and human culture (through socio-cultural evolution). Causes within one sphere may turn into effects in its own sphere, in the other sphere, or in both, which then turn into further causes and further effects. Conceptually limiting events, processes, or forces as having effects only in one sphere ignores the location of human biophysical structures in both natural and cultural spheres of causation. New developments in the cultural representation of nature affect the human colonization of nature by changing culturally acceptable behaviour, resulting in both intentional and unintentional effects. Since each sphere of causation is autonomous and creates new conditions both spontaneously and in response to changes in the other sphere of causation, this becomes a recursive process. This model's explicit acknowledgement of autonomous and spontaneous creativity in both spheres of causation (natural and cultural) helps to prevent circular reasoning, as direct causal relationships are precluded by the terms of the model. Note that the model argues that the pre-condition of any systemic cultural reproduction is the maintenance of its metabolic exchange with its relevant environment.[13]

The interactive model of socio-economic metabolism has particular strengths for environmental historical research. To quote from Hoffmann:

> The interaction model (*sic*) encompasses the dynamic attributes of both nature and human culture and helps pinpoint the kinds of relationships arising in the conjunction. It imparts a temporal dimension to the particular operations of cultural and natural processes while preserving the

autonomy, indeed the mutual indeterminacy, of both causal spheres. As a heuristic device the model provides a means of organizing the evidence of the past to pose and answer relational questions about the interplay of humans and their environment without predetermining those answers.[14]

Relating directly to my research questions, the model indicates that challenges arising from the relevant environment may stimulate developments in the corresponding cultural representation of nature, which could develop as a spontaneous attempt to enhance metabolic opportunities for a society's members in response to stresses on biophysical structures. Since social control of the appropriated environment is justified by the cultural system, long-term failure to adequately manage those environmental elements deemed within social boundaries may summon alternatives from within a culturally acceptable range of options in an attempt to re-exert control. Cultural responses to environmental change are spontaneous and not pre-determined in any *a priori* manner.

Some core concepts of this research now need to be defined; what do I mean by *weather, climate, nature, environment, and culture*? The simplest of these terms is *weather*, although it is a complex reality: *weather* is the immediate condition of the atmosphere at any particular time and place. The components which are present, such as sunshine, cloudiness, temperature, precipitation, humidity/aridity, air pressure, and wind (direction and speed), assemble in various combinations which change over time. The weather may change quickly or slowly, hourly or over days. Experience and/or direct observation, whether measured or unmeasured, generally provides the means of learning about the weather.[15]

Climate, however, is a more abstract concept; understanding it demands lengthier and more sophisticated effort. As described by Mike Hulme in his 2017 monograph *Weathered: Cultures of Climate*, klima was understood differently by the ancient Jewish community and Classical Greek intellectuals, even though it has etymological roots in both. Hulme helps us understand that it is a complex and versatile term with connotations which both describe the climate (for example, when the term is used to describe large-scale physical systems) and assign agency to it (when explaining an array of physical, social, and human outcomes such as species extinctions or wars). Although the term may have been known to theologians of early sixteenth-century Alsace, I was unable to find it in the sources I examined (more about this below).[16]

As a result of that absence, throughout this work I shall use the narrow scientific definition of climate as proposed by the International Panel on Climate Change:

"… climate is usually defined as the average weather, or more rigorously as the statistical description in terms of the mean and variability of relevant quantities over a period of time ranging from months to thousands or millions of years. The classical period for averaging these variables is 30 years, as defined by the World Meteorological Organization (WMO).[17]

The broad definition of climate – the Earth's climate system comprising the atmosphere, the hydrosphere, the cryosphere, the lithosphere, and the biosphere – operates at a scale too large for this discussion of a single bioregion. Further discussion of the climate may be found in Chapter 1, including historical climatology and its findings for Alsace during the late fifteenth and early sixteenth centuries.

The relationship between weather and climate at its simplest is one of raw data to collated averages. The raw data are the weather events which are observed, experienced, and, perhaps, measured on a regular basis, while climate is the accumulation of those weather events into statistically meaningful averages. An analogous relationship would be the relationship of an individual woman's lifespan to national life expectancy rates for women. As individual lifespans may be shorter or longer than national life expectancy, so weather events may be hotter or colder, wetter or dryer, than climate norms for any specific region or period. However, over time, repeated extreme weather events characterized by a single factor (for example, early frosts) will affect the climate in the correlating direction (in relation to the last example, towards cooler averages for the autumn).

The roles of weather and climate in our experience of nature/the environment are central, but distinct. The components of weather determine the quality and quantity of plant, animal, and human life, among other things, in any particular location and period. All the factors of the climate system, listed above, combine to support life on earth. Thus, among other factors and to put it very simply, familiar features of the natural environment at any time and place are contingent upon the specific weather regime of the area, which is itself circumscribed by the possibilities available in the climate system. Should those possibilities change, whether due to diminished or increased solar activity, volcanic activity on earth, anthropogenic greenhouse gases, or developments in the climate system itself, weather regimes will change, different weather events will become the norm for specific areas, and the parameters for plant and animal life in that area will shift, leading to changes in the types and/or quantities of plants and animals which are able to flourish there.[18]

Defining the concepts of *nature* and *environment* is a more intricate challenge than either of the two previous terms. Developments in early sixteenth-century conceptualizations and representations of nature are central to the research questions of this book, and distinguishing them from the twenty-first-century definitions of nature and of the environment will support the unfolding of a coherent argument. At the core of current usage of the term *nature* are the phenomena of the ecologically active physical world collectively, especially the plants, animals, and other features and processes of the earth itself (including both weather and climate). Another important meaning of the term refers to the essential qualities, physical or otherwise, which are vital for identification and knowledge. The difference between the two connotations may be seen in the phrase *the nature of nature*, where the first iteration refers to the essential qualities of the second. The relationship between these two meanings is interesting and worthy of further study. Although many sixteenth-century Alsatians used *nature* to mean *essence*, it is the second signification – the ecologically active physical world, or, as medieval people would say, *Natura* – which I use here.

Only in the second part of the twentieth century did the *environment* come to mean the natural world in general, either as a whole or within a particular geographical area (especially as affected by human activity); the roots of the term are in thirteenth-century Anglo-Norman terms for 'proximity' and 'surroundings.' The word is sometimes qualified, leading to particular renderings such as social environment, aquatic environment, or biotic environment. However, my work here will follow common usage to mean the natural world where it is 'relatively unchanged or undisturbed by human culture.' While the separation is both philosophically and ecologically contentious, considering humanity and its artefacts as distinct from both *nature* and *the environment* is recognized as scientifically acceptable. *Environment* will be used interchangeably with *nature* in this book unless otherwise qualified.[19]

Also important here is a definition of the term *culture*. An interpretative paradigm is adopted here where the goal is to uncover and interpret culture through the context where it exists. Leading proponents of this approach were the American anthropologists Clifford Geertz and Edward T. Hall. The main assumptions shared by those who adopt this perspective is that culture cannot be reduced to an abstract entity, but exists and emerges through details, actions, meanings, and relationships. Communicative behaviours, along with their meanings, are held to constitute culture. At the same time, such behaviours are informed by the culture in which they occur; meaningful communication requires a shared understanding of concepts. From this perspective, the researcher's role is to describe communicative behaviours in detail and in their contexts, as a way of interpreting the culture in its entirety.[20]

The concepts defined earlier in this section may now be placed in relation to the interactive and hybrid model of socio-economic metabolism and their role in this volume clearly articulated. That is: weather, in the short term, and climate, in the long term, are central determining factors (although not the sole determining factors) in the quality and quantity of plants, animals, and other ecologically relevant elements which, collectively, constitute nature. Due to the unavoidably material aspects of our lives, human societies are embedded in nature and depend upon it for existence; the biophysical aspects of human society are subject to natural laws. The effect of the weather and climate on nature is therefore important to human societies for both specific and general reasons.

Specifically, weather plays an influential role in determining, among other things, human health, crops, energy resources, the behaviour and health of animals (wild and domestic), population levels, and, ultimately, economic and social stability. Climate, whether understood narrowly or broadly, determines the general weather regime for any particular period and place, upon which successful subsistence patterns and the general economic expectations and structures of a society are founded. Accordingly, I provide a report of recent historical climatological research on the climatic context of the area in Chapter 1; Chapters 2–4 present annual weather reports for 1473 to 1509, 1509 to 1525, and 1525 to 1540, respectively, highlighting any notable weather events which may have affected the region. Each section of the weather report includes some of the social consequences of the weather, such as economic and social instability as well as renovations of Strasbourg's welfare system. However, in this book I give primary attention to the articulation of Christian representations of nature; I consider these to be spontaneous developments in the cultural sphere of causation arising in response to stress from the natural sphere of causation.

It would be very difficult to deny entirely that the natural environment has an influence upon any culture's concept of *nature*. As articulated in the model of socio-economic metabolism, since culture is a sphere of autonomous and unlimited creativity by human beings, the natural environment does not hold a direct and causative role in cultural representations of nature. However, such representations germinate, develop, and diminish in social contexts where, among other things, existing cultural values, political conflict, and economic imperatives exert influence, along with any particular natural environment. Entailments of one concept may be applied to other concepts; socially appealing conceptual developments may be more widely shared than those less attractive and, quickly or slowly, may become normative and motivate new social behaviour. In this multi-faceted context of cultural development, searching for a single definitive cause for transformations in a culture's concept of *nature* is unrealistic. It is, however, essential to include natural conditions among the fluctuating combination of variables and influences contributing to developments in the concept of *nature*. This is due to their central role in establishing the environmental context in which conceptual developments occur (particularly relevant when the concept in question is nature itself) and their contributions to the social and economic contexts of the same.

In keeping with the hybrid, interactive model of socio-economic metabolism, then, my research describes the natural sphere of causation around Strasbourg, with particular attention to the weather and climate from 1473 to 1541; discusses its effects on Alsatian society during the last decades of the fifteenth century and the first four decades of the sixteenth century; and explores Catholic, radical, and Reform views of nature from 1509 to 1539. Juxtaposing the natural, social, and cultural spheres not only demonstrates the speed with which culturally acceptable views of nature may change; it also permits an exploration of the possibility that social and economic instability in the bioregion (in part weather related) supported the appearance or acceptance of new religious representations of nature among theologians in Strasbourg. This is a good moment to offer a reminder that, since cultural developments are spontaneously created by human beings (even if they are responsive to stress from the natural sphere), conclusions about causation cannot be drawn.

Temporal and geographic parameters

Climate historians study the interplay of weather and climate with society and culture. As described by Reinhold Reith in his volume *Umweltgeschichte der Frühen Neuzeit*, though, the challenges of time and space are keen for environmental historians, and only more so for climate historians. In the wider sub-field of environmental history, the division between solar-powered societies and those relying upon fossil fuels provides a useful distinction between agricultural and industrial periods; changes in other human behaviour in the environment, such as those related to mining, agriculture, or forestry, also allow for the delimitation of ages and periods in research.[21]

My research accepts a broad understanding of the Little Ice Age (LIA) as a primary determinant for the period of study, taken to have begun in the late thirteenth and early fourteenth centuries as determined by a combination of decreased solar activity, growth of the Atlantic ice pack, radiocarbon dating, cessation of reliably warm summers in Northern Europe, and the rains which led to the Great Famine of 1315–1317, among other factors. During the LIA, atmospheric conditions were not uniform and reflected variance in solar activity influenced by terrestrial events like volcanoes. Climatologists have identified peaks and troughs of temperature averages during the LIA, such as the Wolf Minimum (c. 1280–1340), the Spörer Minimum (c. 1420–1570), and the Maunder Minimum (c. 1645–1715). My weather report extends from 1473 to 1540, in the middle portion of the Spörer Minimum.[22]

Examining the role of weather within the many cultures and societies of early modern Europe would be the work of a lifetime. To achieve a meaningful analysis of the impacts of climate on culture within the limitations of a monograph, my spatial focus necessarily narrows to the Upper Rhine Valley, chosen for a combination of reasons. Primarily, the Upper Rhine Valley is a geographically unified area, a tidy bioregion which was recognized by sixteenth-century Europeans as *Elsass* (Alsace). It was also a distinctive cultural region with its own dialect, bordered by France, German Swabia, and the Swiss Confederacy. The geographical and cultural boundaries of this project are generally synonymous with those of the Upper Rhine Valley; the semi-continental climate and other material conditions of the region will be described in Chapter 1.

Another reason for focusing on the Upper Rhine Valley is the wealth of secondary information available on the people, culture, and social developments of Strasbourg. A clear debt is due to scholarly giants Thomas A. Brady, Jr., Georges Bischoff, Rita Voltmer, Miriam Usher Chrisman, Tom Scott, and others upon whose work my investigation builds. Much of their research focuses tightly on Strasbourg, and my exploration, as well, emphasizes the city and its role in cultural developments. Although important social events like the Peasants' War or the Reformation occurred as much in the countryside as the city, or among both rural and urban populations together, it is fruitless to deny the central role played by Strasbourg in the history of Alsace.

A benefit of drawing relatively tight geographical limitations is that weather conditions are similar throughout the valley, leading to common expectations among Alsatians about what constituted good or bad weather, a familiar knowledge of the effect of particular weather events, and shared hopes for seasonal atmospheric behaviour and outcomes. Such shared views and knowledge are the basis of cultural perceptions, as while individual weather events could have been as localized as a ten-minute rain shower over Colmar or as continental as the heat wave of 1473, the cultural context within which weather events occurred was a regional one, with shared expectations, knowledge, and linguistic references from Speyer to Basel. A note of caution, though: geographic limitations accepted for this research project do not imply cultural isolation. There were profound similarities (linguistic, religious, legal, and financial, among others) with neighbours north and east of the Upper Rhine Valley, and financial and legal ties to the west and southwest, as well as distinctions between the communities.

Returning to a discussion of temporal parameters, this time within the cultural sphere of causation, my decision to focus on printed material published between 1509 and 1539 is the consequence of an intention to emphasize changes introduced to Alsatian cultural norms by different religious perspectives. Developments in views of the natural world may occur organically over generations, but the sharp theological differences expressed by Roman Catholics, radical peasants, and Lutheran Evangelicals, among others, openly contested for influence through popular appeal for only a handful of years. By 1540, institutional authority had overcome individual religious inspiration throughout the region, and orthodoxy was once more enforced by civil law. Although theological differences between Catholicism, Lutheranism, and Anabaptism continued to develop, many of the roots of their divergence may be found in the claims and disputes of the early sixteenth century.

Looming over the medieval, early modern, and later centuries, the LIA is only beginning to be taken into consideration by historians interested in these periods. Midway through the LIA, German reformers successfully challenged the cultural authority of the Roman Catholic Church, but failed to establish a more equitable social order despite widespread rebellion by peasants in southern Germany. These challenges took place during the Spörer Minimum, a trough of the Little Ice Age featuring 90 years of unreliable and unstable weather conditions and characterized by an increased numbers of extreme weather events which placed heavy pressure upon the agricultural society of Europe. This junction of a heightened frequency of extreme weather events with social and cultural turmoil inspires my research questions.

Translation of nomenclature and calendars

The dominant language in many of the locales included in this research has changed during the five centuries since these events took place, often several times. The traditional language of the Upper Rhine Valley in the early sixteenth century was Alsatian, one of a group of Low Alemannic German dialects spoken in the southwestern corner of the German speaking lands. The shared literary languages of the region were Latin and what is now referred to as Alemannic Middle High German (*Alemannischmittelhochdeutsch*). In an effort to avoid confusion while remaining faithful to original place names, I include in parentheses translations when they are first introduced (i.e. *der Bodensee* (Lake Constance). The name of a person will be presented as they may have introduced themselves, untranslated from their native tongue.

The Julian calendar was in use throughout the period studied for this research. In order to be faithful both to my primary sources and to the natural world, which is more accurately measured by the Gregorian calendar, I judge it most appropriate to provide both dates when needed, rather than prioritize one calendar over the other. Therefore, I provide first the Julian date, followed by a slash, and then the Gregorian date (i.e. *1/10 January* or *31 March/9 April*). Although an adjustment of nine days would be required for dates in the fifteenth century, to avoid confusion I base non-specific Gregorian equivalents, such as saints' days, on the ten-day adjustment appropriate to the sixteenth century. If there is not a specific calendar date available, as when sources mention only a month, I will remain with the Julian month mentioned.[23]

Research goals

As the most volatile element of the natural world during the research period, what role (if any) did the weather play in social and cultural turbulence in Strasbourg and Alsace from 1509 to 1541? Were there similarities or differences in representations of the natural world among Strasbourg's Catholic, radical, and Evangelical theologians? If so, what were they and did they contribute to change in the approach of Strasbourgeois and/or Alsatians to the environment? These research questions inspire an exploration of differences in the religious representation of nature and Christian approaches to nature in Strasbourg during the triple junction of unreliable weather conditions characteristic of the Spörer Minimum, the cultural changes introduced by early religious reform, and the social turmoil of the 1525 German Peasants' War.

Much research has been conducted on the Reformation, on the 1525 German Peasants' War, on individual theologians featured in this project (Geiler and Capito, in particular), and on the demographic and economic history of Alsace, as well as upon reconstructions of the weather and climate during the early sixteenth century. To date, however, no research has explored potential relationships between the heightened number of extreme weather events and social strife, economic problems, or developments in the cultural representation of nature.

Sixteenth-century Christendom was not homogeneous, and searching for evidence of the role played by weather in cultural change or social unrest requires establishing tight temporal and geographic parameters in order to be confident that individuals whose work may be juxtaposed were likely to have experienced the same overall weather conditions. Through three case studies, my research queries Strasbourg theologians from 1509 to 1539 for the manner in which they represented nature. I describe six decades of weather in Alsace, outline its effects upon Strasbourg and Alsatian society, and explore three representations of nature made by Roman Catholic, radical, and Evangelical theologians, and their views of a suitably Christian relationship to the natural world. In particular, the sources are examined for evidence of continuity or rupture in the religious representation of the environment.

The tight time frame adopted in my research allows for an in-depth examination of sources, although scarcity prevents study of a single genre – a particular disadvantage for research comparing views of nature across religious difference. For example, although sermons were regularly published in the period, those of Strasbourg's reformers are surprisingly absent from the archive; instead, publication of polemical tracts, Scriptural exegesis, letters, or other genres gained in popularity and are amply represented. However, despite differences in genre of presentation, the authors of the selected sources shared a common goal and a common method in achieving that goal: to persuade their audience to adopt a more virtuous life, each author appealed to Scripture and developed his own exegesis within a single overarching frame, that provided by Christianity. Although using material from different genres is a not-insignificant drawback, this research proceeds on the assumption that commonalities among the authors of the primary material – that they shared goals, methods, references, a time frame, and a single natural environment – will outweigh the drawbacks of different genres of source material.

My research into cultural change, then, juxtaposes an annual description of weather events with some later sermons of Roman Catholic preacher Johannes Geiler von Kaysersberg, new views introduced by radical religious preacher Clemens Zyegler and the peasant authors of the *XII Artikels*, and Wolfgang Capito's exegesis of the first book of Genesis, *Hexemeron Dei opus*. My choice of primary sources takes advantage of hindsight in choosing the authors, prioritizing cultural leadership and/or popularity in selecting these specific works, attributes to which the authors may have aspired but could not be certain of achieving.

I focus upon cultural representations of nature, and while the environmental frame will be described in some detail, the choice of primary source material was made on the basis of three criteria: that it was published in the location and time frame mentioned above; that the author understood himself and was recognized by others to write from one of the religious perspectives attempting to maintain or achieve interpretative authority within the culture; and that the author enjoyed a high degree of influence within his community. The analyses will be ordered in the chronological sequence in which they were published.

Johannes Geiler von Kaysersberg's collection of sermons published as *Die Emeis* was an obvious choice to represent views from the Roman Catholic community in Strasbourg and will be the subject of Chapter 2. Geiler's education as a Doctor of Theology, the position he occupied for over three decades as the cathedral preacher in Strasbourg's *Liebfrauenmünster*, the privileged status he enjoyed in his community, and his position as chaplain to Maximilian I, Emperor of the Holy Roman Empire, render his voice uniquely influential.

Two selections were made from the peasant community, the focus of Chapter 3: a pamphlet written by radical Strasbourg gardener Clemens Zyegler entitled *Ein Fast Schon Büchlin* and the peasant manifesto *XII Artikels*. Zyegler, although self-taught, was an influential exponent of radical religious views, as demonstrated by the invitation he received to preach to peasant rebels near Mont Ste-Odile in 1525. The radical nature of his theological perspective was confirmed by the Strasbourg Evangelical reformers, who acknowledged his prominence when allocating him first place as a debating opponent in the synod they organized to define Strasbourg's municipal orthodoxy in 1533. The *XII Artikels*, collectively authored by a committee of 50 rebels and edited by a lay theologian and an Evangelical pastor, were carried throughout the southwestern German-speaking lands by rebelling peasants.

The views of Reform theologian Wolfgang Capito contributed to establishing new standards of religious orthodoxy in Strasbourg. Like Geiler, Capito was a Doctor of Theology; from 1519 to 1523, he advised Albrecht von Brandenburg (1490–1545), the young Elector and Archbishop of Mainz and Magdeburg, Prelate of Germany. Upon his move to Strasbourg in 1523 and ensuing acceptance of Evangelical views, Capito's status established him as one of four leading reformers in the Free Imperial City. The *Hexemeron Dei opus* was his last publication, and likely was written as a textbook for the students of Johannes Sturm's *Schola Argentoratensis*, which had opened the year before it was printed. It will be discussed in Chapter 4.

The challenge with limiting primary sources to a specific bioregion and time period is that it makes it difficult to generalize from these particular research findings to a broader historical period. However, the climatic and cultural conditions which were characteristic of early sixteenth-century Alsace were not physically or culturally estranged nor distant from those in neighbouring regions of the Holy Roman Empire and the Swiss Confederacy, as shown by the continental nature of some extreme weather events, the trade in grain between neighbouring regions, Alsatian peasants joining the Swabian peasants in the 1525 uprising, and the high levels of communication between Strasbourg's Reformers and Luther and his colleagues in Württemberg. While these research findings are specific to Alsace, there are reasons to expect that they may be applicable to other areas where the Reform succeeded.

Every text works to position readers in a specific way and asking questions of early sixteenth-century texts whose answers will be meaningful to present issues means distancing oneself from the intentions of the authors, 'reading against the text.' My research is situated within the growing wave of environmental humanities; as difficult to pin down epistemologically as women's studies, the field of environmental humanities is characterized as a body of study which 'simultaneously and holistically examines issues of knowledge and power in light of a culturally (socially, historically) constituted nature and an environmentally constituted culture.' My research queries the selected texts in frames which would not necessarily be recognizable to the authors and prioritizes concerns which were not theirs. However, to limit the present inquiry to a framework based on the views and understandings of the early sixteenth century would severely and unnecessarily limit the insights that may be gained.[24]

Despite the desires of philosophers, technocrats, and bureaucrats, human beings generally make decisions based on what feels right and true to them. Explanations and guidance for 'right' and 'true' are more often found in religious faith than in intellectual effort, particularly in periods when the worldview that permeates and dominates a culture is a religious one (unlike the economic worldview which holds sway over our twenty-first-century Western culture). Asking questions about the concept, representation of, and approach to the natural world during the European early modern period therefore draws this historian's attention to the views of religious leaders.

Notes

1 Max Weber, 'Science as a Vocation,' in *From Max Weber: Essays in Sociology*, trans. and ed. Hans Heinrich Gerth and C. Wright Mills (New York: Oxford University Press, 1946), 139; first published in an edited edition as "Wissenschaft als Beruf," in *Gesammelte Aufsätze zur Wissenschaftslehre* (Tübingen: Mohr (Siebeck), 1922), 524–55; originally a speech at Munich University, 1918, published in 1919 by Duncker & Humboldt, Munich.
2 Robert Scribner, 'The Reformation, Popular Magic, and the "Disenchantment of the World,"' *Journal of Interdisciplinary History*, 23 (1993), 475–94.
3 Tom Scott, *The Early Reformation in Germany: Between Secular Impact and Radical Vision* (Farnham: Ashgate, 2013), 15.
4 Alexandra Walsham, 'The Reformation and 'The Disenchantment of the World' Reassessed',' *The Historical Journal*, 51 (2008), 497–528.
5 See, for example, Keith Thomas, *Religion and the Decline of Magic: Studies in Popular Beliefs in Sixteenth and Seventeenth Century England* (Oxford: Oxford University Press, 1971) and Keith Thomas, *Man and the Natural World: Changing Attitudes in England 1500–1800* (Oxford: Oxford University Press, 1983).

6 Richard Kieckhefer, 'The Specific Rationality of Medieval Magic',' *The American History Review*, 99 (1994), 813–36; Edward Grant, *The Nature of Natural Philosophy in the Late Middle Ages* (Washington: The Catholic University Press of North America, 2010); Stuart Clark, *Thinking with Demons: The Idea of Witchcraft in Early Modern Europe* (Oxford: Oxford University Press, 1997).

7 Lynn White, Jr., 'The Historical Roots of Our Ecological Crisis,' *Science*, n.s., 155:3767 (1967), 1203–7 (1205).

8 Wolfgang Behringer, 'Climate Change and Witch-Hunting: The Impact of the Little Ice Age on Mentalities',' *Climatic Change*, 43 (1999), 335–51; Wolfgang Behringer, *Kulturgeschichte des Klimas* (Munich: C.H. Beck oHG, 2007); translated into English and published as *A Cultural History of Climate* (Cambridge: Polity Press, 2010).

9 John L. Brooke, *Climate Change and the Course of Global History: A Rough Journey* (Cambridge University Press, 2014); Benjamin Lieberman and Elizabeth Gorden, *Climate Change in Human History: Prehistory to the Present* (Bloomsbury Academic, 2018); Sam White, Christian Pfister, and Franz Mauelshagen, eds., *The Palgrave Handbook of Climate History* (London: Palgrave MacMillan, 2018); Christan Pfister and Heinz Wanner, *Climate and Society in Europe – The Last Thousand Years* (Haupt Verlag, 2021); Mike Hulme, *Weathered: Cultures of Climate* (London: SAGE Publications, 2017), 4. Note that in considering Alsatians' knowledge of climate to be indigenous, I am not claiming that Alsatians are an Indigenous people as understood by the 2007 United Nations Declaration on the Rights of Indigenous Peoples.

10 Thomas Kirchhoff and Ludwig Trepl, eds., *Vieldeutige Natur: Landschaft, Wildnis und Ökosystem als kulturgeschichtliche Phänomene* (Bielefeld: Transcript, 2009), 14.

11 1Marina Fischer-Kowalski and Helga Weisz, 'Society as Hybrid Between Material and Symbolic Realms: Toward a Theoretical Framework of Society-Nature Interaction,' *Advances in Human Ecology*, 8 (1999), 215–51.

12 Fischer-Kowalski and Weisz, 228–30; Richard Hoffmann, *An Environmental History of Medieval Europe* (Cambridge: Cambridge University Press, 2014), 8.

13 Hoffmann, 10; For further discussion of the perils of circular reasoning in historical climatology, see Franz Mauelshagen, *Klimageschichte der Neuzeit, 1500–1900* (Darmstadt: Wissenschaftliche Buchgesellschaft, 2010), 41.

14 Hoffmann, 10.

15 Mauelshagen, 6.

16 Hulme, 15–17.

17 The quantities mentioned are variables at the earth's surface, such as temperature, precipitation, and wind. Intergovernmental Panel on Climate Change, 'Glossary,' <https://apps.ipcc.ch/glossary/> [accessed 3 October 2023].

18 A weather regime is a recurrent large-scale spatial atmospheric structure with a deformation radius of at least several hundreds of kilometres, usually described in terms of circulation variables (air pressure, geopotential height, etc). Mathieu Vrac and Pascal Yiou, 'Weather Regimes Designed for Local Precipitation Modeling: Application to the Mediterranean Basin,' *Journal of Geophysical Research*, 115 (2010), D12103. doi: 10.1029/2009JD012871, 1.

19 Oxford English Dictionary online, 'environment,' <https://www.oed.com/dictionary/environment_n?tab=etymology> [accessed 5 October 2023]; Donald Lee Johnson et al., 'Meanings of Environmental Terms,' *Journal of Environmental Quality*, 26 (1997), 582.

20 Zhu Hua, 'Identifying Research Paradigms,' in *Research Methods in Intercultural Communication: A Practical Guide*, ed. Zhu Hua (Chichester: Wiley-Blackwell, 2016), 8.

21 Reinhold Reith, *Umweltgeschichte der Frühen Neuzeit*, Enzyklopädie Deutscher Geschichte 89 (Munich: Oldenbourg, 2011), 4–9.
22 Fredrik Charpentier Ljungqvist, 'A New Reconstruction of Temperature Variability in the Extra-Tropical Northern Hemisphere during the Last Two Millennia,' *Geografiska Annaler*, Series A, Physical Geography, 92 (September 2010), 338–51; see also Rudolf Brázdil et al., 'Historical Climatology in Europe – The State of the Art,' *Climatic Change*, 70 (2005), 388–92; Mauelshagen, 60; Ljungqvist, 346.
23 For efficient conversion of dates, I have taken advantage of Fourmilab Switzerland's calendar converter, available here: http://www.fourmilab.ch/documents/calendar/#juliancalendar.
24 Rich Hutchings, 'Understanding of and Vision for the Environmental Humanities,' *Environmental Humanities*, 4 (2014), 216.

References

Secondary sources

Behringer, Wolfgang, 'Climate Change and Witch-Hunting: The Impact of the Little Ice Age on Mentalities,' *Climatic Change*, 43 (1999), 335–51.
——— *Kulturgeschichte des Klimas* (Munich: C.H. Beck oHG, 2007); translated into English and published as *A Cultural History of Climate* (Cambridge: Polity Press, 2010).
Brázdil, Rudolf, Christian Pfister, Heinz Wanner, Hans von Storch, and Jürg Luterbacher, 'Historical Climatology in Europe – The State of the Art,' *Climatic Change*, 70 (2005), 388–92.
Brooke, John L., *Climate Change and the Course of Global History: A Rough Journey* (Cambridge: Cambridge University Press, 2014).
Clark, Stuart, *Thinking with Demons: The Idea of Witchcraft in Early Modern Europe* (Oxford: Oxford University Press, 1997).
Fischer-Kowalski, Marina, and Helga Weisz, 'Society as Hybrid Between Material and Symbolic Realms: Toward a Theoretical Framework of Society-Nature Interaction,' *Advances in Human Ecology*, 8 (1999), 215–51.
Fischer-Kowalski, Marina, Helmut Haberl, Fridolin Krausmann, Verena Winiwarter, eds., *Social Ecology: Society-Nature Relations Across Time and Space* (Switzerland: Springer Internaional Publishing, 2016).
Grant, Edward, *The Nature of Natural Philosophy in the Late Middle Ages* (Washington: The Catholic University Press of North America, 2010).
Hoffmann, Richard C., *An Environmental History of Medieval Europe* (Cambridge: Cambridge University Press, 2014).
Hulme, Mike, *Weathered: Cultures of Climate* (London: SAGE Publications Ltd., 2017).
Hutchings, Rich, 'Understanding of and Vision for the Environmental Humanities,' *Environmental Humanities*, 4 (2014), 213–20.
Johnson, Donald Lee, S.H. Ambrose, Thomas Bassett, and M.L. Bowen, 'Meanings of Environmental Terms,' *Journal of Environmental Quality*, 26 (1997), 581–9.
Kieckhefer, Richard, 'The Specific Rationality of Medieval Magic,' *The American History Review*, 99 (1994), 813–36.
Kirchhoff, Thomas, and Ludwig Trepl, eds., *Vieldeutige Natur: Landschaft, Wildnis und Ökosystem als kulturgeschichtliche Phänomene* (Bielefeld: Transcript, 2009).

Lieberman, Benjamin, and Elizabeth Gorden, *Climate Change in Human History: Prehistory to the Present* (London and New York: Bloomsbury Academic, 2018).

Lungqvist, Fredrik Charpentier, 'A New Reconstruction of Temperature Variability in the Extra-Tropical Northern Hemisphere during the Last Two Millennia,' *Geografiska Annaler*, Series A, Physical Geography, 92 (2010), 338–51.

Mauelshagen, Franz, *Klimageschichte der Neuzeit, 1500–1900* (Darmstadt: Wissenschaftliche Buchgesellschaft, 2010).

Pfister, Christian, and Heinz Wanner, *Climate and Society in Europe – The Last Thousand Years* (Berne: Haupt, 2021).

Reinhold Reith, *Umweltgeschichte der Frühen Neuzeit*, Enzyklopädie Deutscher Geschichte 89 (Munich: Oldenbourg, 2011), 4–9.

Scott, Tom, *The Early Reformation in Germany: Between Secular Impact and Radical Vision* (Farnham: Ashgate, 2013).

Scribner, Robert, 'The Reformation, Popular Magic, and the "Disenchantment of the World,"' *Journal of Interdisciplinary History*, 23 (1993), 475–94.

Thomas, Keith, *Religion and the Decline of Magic: Studies in Popular Beliefs in Sixteenth and Seventeenth Century England* (Oxford: Oxford University Press, 1971).

—— *Man and the Natural World: Changing Attitudes in England 1500–1800* (Oxford: Oxford University Press, 1983).

Vrac, Mathieu, and Pascal Yiou, 'Weather Regimes Designed for Local Precipitation Modeling: Application to the Mediterranean Basin,' *Journal of Geophysical Research*, 115 (2010), D12103. doi: 10.1029/2009JD012871.

Walsham, Alexandra, 'The Reformation and 'The Disenchantment of the World' Reassessed,' *The Historical Journal*, 51 (2008), 497–528.

Weber, Max, 'Science as a Vocation,' in *From Max Weber: Essays in Sociology*, trans. and ed. Hans Heinrich Gerth and C. Wright Mills (New York: Oxford University Press, 1946), 139; first published in an edited edition as 'Wissenschaft als Beruf,' in *Gesammelte Aufsätze zur Wissenschaftslehre* (Tübingen: Mohr (Siebeck), 1922), 524–55.

White, Lynn, Jr., 'The Historical Roots of Our Ecological Crisis,' *Science*, n.s., 155:3767 (1967), 1203–7.

White, Sam, Christian Pfister, and Franz Mauelshagen, eds., *The Palgrave Handbook of Climate History* (London: Palgrave MacMillan, 2018).

Zhu Hua, 'Identifying Research Paradigms,' in *Research Methods in Intercultural Communication: A Practical Guide*, ed. Zhu Hua (Chichester: Wiley-Blackwell, 2016), 3–22.

Digital sources

Fourmilab Switzerland, <http://www.fourmilab.ch/documents/calendar/#juliancalendar>.
Intergovernmental Panel on Climate Change, 'Glossary,' <https://apps.ipcc.ch/glossary/>.
Oxford English Dictionary online, <https://www.oed.com/dictionary>.

1 Physical and social frames

The Upper Rhine Valley and pre-reform Alsatian society

Introduction

Any attempt to comprehend the influence of Geiler, Zyegler, and Capito's representations of nature leads to questions not only about their ideas and thoughts, but also about their audiences and the natural surrounds in which they all lived. What is the Upper Rhine graben, politically identified as Alsace in 1509? What type of weather was ideal for this agriculturally dominated region? Who stood in the Cathedral to listen to Geiler's sermons or in the fields to find inspiration from Zyegler? Who took notes on Capito's lectures? What were the material conditions and social expectations of these people? How were their lives ordered by social, political, and economic relationships and institutions and how did those relationships and institutions respond when they came under stress from the natural environment?

The hybrid and interactive model of socio-economic metabolism suggests that, in order to understand the cultural responses of early sixteenth-century Alsatians and Strasbourgeois to the weather events of their time, historians must look first to the natural environment of Alsace and then to the social structures, including infrastructure, behaviours, and tensions which serve as bridges between the natural and cultural spheres of causation. That is, from a very general perspective, most human social behaviour takes place within a meaningful cultural framework and before we can fruitfully juxtapose environmental stress with changes in the cultural frame, it is useful to describe the environment and society within which competing cultural perspectives were expressed.

Chapter 1, then, builds a geophysical portrait of the Upper Rhine graben, including brief descriptions of the geography, Rhine and Ill Rivers, forests, vineyards, and agriculture, and provides a profile of the human society in late medieval Strasbourg and surrounding Alsace against which later changes and events may be understood. The chapter begins with an outline of the natural environment with as much fidelity as possible to the period in which the authors of my case studies lived, interwoven with information about Alsatians' engagement with their natural world, as appropriate. I will also discuss the methods of weather reconstruction as developed in the sub-field of historical

DOI: 10.4324/9781003262411-2

climatology, explain why such a weather reconstruction is not possible for Strasbourg and the surrounding area at this time, and introduce a recent weather reconstruction centred on nearby Basel. I also introduce the sources used for my weather reports in Chapters 2–4.

The chapter goes on to feature a summary of late medieval institutional structures, social relations, and social tensions in the city and region, organized around an expansion of Geiler's imagined potential audience for sermons in the *Liebfrauenmünster zu Straßburg*. Along with establishing the Alsatian geography, my immediate aim is to approximate the possible range of audience members and, thereby, the potential reach of influence exerted by his Catholic representations of the natural environment. A secondary aim is to situate Geiler within his community because it was vulnerable demographically and economically to extreme weather events and an understanding of the consequences of these events relies on some knowledge of how people organized their social relations. Thankfully, substantial previous research permits an outline of social conditions and institutional structures of the city and region.[1]

Physical geography

In the twenty-first century, the Upper Rhine River from Basel to Lauterberg, a few kilometres southwest of Karlsruhe, functions as a border between the nation-states of France and Germany. We have become accustomed to several hundred years of a story where Alsace features as a militarily 'Contested Land,' being held as German territory until 1648, French territory from 1648 to 1870, German from 1870 to 1918, French from 1918 to 1939, German from 1939 to 1945, and French lands since 1945. Other features of the landscape, however, call for recognition of the Upper Rhine Valley's identity as a single unit. That the fifteenth- and sixteenth-century Alsatian felt herself or himself to be living in a coherent region was made explicit by Sebastian Münster's 1544 *Cosmographia*, which sharply distinguished Alsace from France, Germany, and Switzerland. Historians of the early modern period regularly accept the bioregional understanding of their subjects; Georges Bischoff examines the *Bundschuh* revolutions on this basis, and political economist Tom Scott conducts his analysis from this same perspective. As well as constituting the natural sphere of causation in the interactive model of socio-economic metabolism for this research, the region's unity was assumed by contemporaries and is accepted by historians of the area; the geographical features which combine to create the bioregion of Alsace, therefore, merit attention.[2]

Although the focus of this book is on Strasbourgeois' religious representations of nature, the interactive model of socio-economic metabolism asserts that spontaneous activity in the autonomous cultural sphere may occur in an indirect relationship to events in the natural world, mediated through the human biophysical experience of the natural environment. That is, that unpredictable changes in human culture may occur in response to events in the natural world (as well as events and developments in the cultural sphere itself) when natural events have an effect on social conditions and behaviour. An understanding of the Alsatian physical environment of the late fifteenth and early sixteenth centuries is therefore important for understanding the role played by that environment in Alsatian social events and for suggesting an indirect explanation for the different representations of nature found in the distinct conceptual packages of Catholic, Evangelical, and Radical theologians as delivered within 30 years of each other.

The Upper Rhine Valley is a shallow river plain flanked by mountains, and this geological formation influences Alsace's temperature, cloudiness, precipitation, evaporation, and energy budget. Their interaction shapes and influences such local weather phenomena as storm patterns, wind velocity and direction, and the likelihood of frost, among other things. The physical and chemical environments established by the basic geology of a region exert a clear effect on the variety and volume of plant life it is able to support. Establishing the entire interactive and reciprocal relationship between humans and nature in Alsace includes geology in a description of the natural environment.[3]

Technically, the Upper Rhine Valley is a trench with a width of 30–40 km and a length of approximately 300 km, interrupting the terrain on a roughly NNE/SSW axis from Basel in the Swiss Alps to Frankfurt-am-Main in the German state of Hesse (see Map 1.1). The trench is of a type known geologically as a *graben*. That is, depressed ground detached from its surroundings by two parallel faults, where the land outside the faults has lifted to become a *horst*. To identify the area in today's terms, on the west, the southerly section of the horst is the Vosges Mountain range in France (*Les Vosges*) and, further north on the same side, the Haardt Mountains in Germany (*der Hardtberge*). The eastern horst, all in Germany, consists of the Black Forest to the south (*der Schwarzwald*) and the Oden Forest to the north (*der Odenwald*). The area of concern for this research is in the southern part of the Upper Rhine graben, between and including the Vosges Mountains and the Black Forest.[4]

The core of the Vosges Mountains is red sandstone; it is found in great abundance in the Northern Vosges (north of Saverne) and regularly in the Upper Vosges (south of Saverne). Its service as a building material was prized, as seen in the use of the stone for Strasbourg's cathedral, the *Liebfrauenmünster zu Strasbourg*. Lead, copper, silver, and iron are present in significant volumes, making the area one of the greatest sources of pre-Columbian silver in Europe. Mining began in the area during the Celtic pre-Roman Iron Age from the fourth century BCE, waxed and waned with the Roman and Alemannic cultures, and was resumed during the tenth century. A particular centre for the extraction of silver was centred on Ste-Marie-les-Mines in the Lièpvre Valley of the Vosges Mountains. After closing down in the early fourteenth century, the industry flourished from 1502, following an investment agreement between the landowner, Guillaume de Ribeaupierre, and Sigismund of Austria.[5]

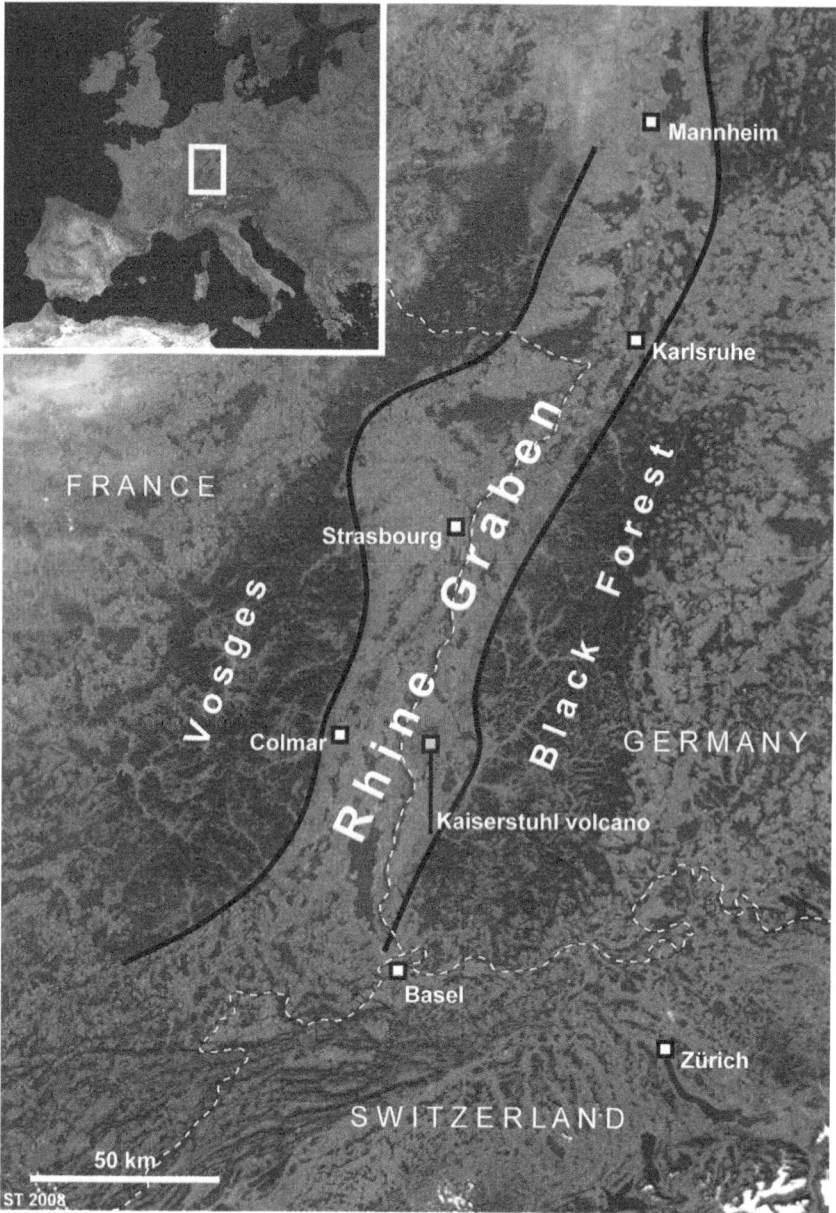

Map 1.1 The Upper Rhine graben.

The Vosges massif rises slowly from the large Paris Basin in north-central France and breaks off sharply in the east; slopes down to the Upper Rhine floor are precipitous. The crest of land separating the watersheds is closer to the Upper Rhine Valley than to the middle of the mountain range. One might get the sense of an impenetrable mountainous wall defining the western rim of the Upper Rhine Valley, but two large passages are provided by the Burgundian Gate separating the Swiss Jura and the southernmost Vosges mountains, and the Severne Pass, which provided reliable, year-round access for medieval travellers between Strasbourg and the nearby cities of Besançon, Dijon, Nancy or Metz. There are as well numerous small and large passes through the Vosges. These include tiny defiles which lead up and out of the Upper Rhine Valley, providing space for footpaths or animal trails, the larger *col de Bussang* linking Basel to the headwaters of the Moselle River, the *col de Bonhomme* linking Colmar to the headwaters of the Meurthe River, and the *col de Saales* running from the valley of the Bruche to Raon-l'Étape or St-Dié, linking Strasbourg with towns to the west of the range.[6]

Likewise the eastern horst of the Upper Rhine graben, known as the Black Forest (*der Schwarzwald*), is home to rivers such as the Rench, the Murg, the Wagensteigbach, the Höllenbach, or the Kinzig which slope down to the Rhine, creating passage for animals and humans as well as water. The Kinzig River valley, for example, was once known as a *Brückenlandschaft*, or 'bridging territory,' for the access it provided from Strasbourg to Schramberg and Rottweil. Numerous smaller passes through the range also existed, such as those of the Simonswälder, the Prech, and the Glotter valleys. Alsace's human and animal populations were not isolated from their neighbours. However, the Black Forest can be seen more as a geological sibling to the Vosges than an identical twin. Where the Vosges Mountains consist mostly of red sandstone, the same is found only on the Enz heights and eastwards; granite and gneiss are proportionally present in greater volumes. Precious and base metals are present, but in smaller volumes than in the Vosges, and mining in the Black Forest was not as important to the region's economy as mining the Vosges.

The southern boundary of the Upper Rhine Valley is established by the Blauen Peaks, which are located at the northeastern limits of the Jura Massif. The ridge stretches westwards into Burgundy to create the southern wall of the Burgundian Gate. Formed from the disrupting fault patterns of the nearby Upper Rhine graben, the Ferrette Jura is considered part of the foreland of the northwestern Alps. Substantially different in composition from its geological neighbours to the north, the Jura consists largely of limestone and clay; mining was not a significant economic activity. In the Jura to the south and east of Basel, the deeply eroded valleys of the Birs, Ergolz, or Frick Rivers do not offer easy passage to the uplands. There is no clear definition of watershed or height of land; settlement was sparse and remains so.[7]

At its northern end, the Upper Rhine graben splits in two and becomes the southern leg of a Y-formation of grabens known as the Rhenish Triple Junction. The branch which leads to the northwest, and through which the Rhine River flows, curls around the Mainz Basin, and then continues through the Rhenish Massif into the Lower Rhine Embayment. The branch which leads to the northeast follows the Hessen depression and the Leine graben near Göttingen, before apparently disappearing beneath the sediments of the North German Plain.[8]

With respect to the climatic conditions of Alsace, the Vosges Mountains in the west create a rain shadow on the western Rhine plane, with more precipitation striking the eastern foothills of the Black Forest. That wetter, western edge of the Black Forest faces the Upper Rhine Valley in a gentler manner than the Vosges, showing greater levels of erosion. The Blauen Peaks, as the northernmost face of the Alps, act as a barrier preventing the passage of warm air from the Mediterranean to the Upper Rhine Valley, while the comparatively gentle opening of the graben to the north allows the moderation of the maritime climate to penetrate further south than at similar latitudes further west.

A further effect of its geological foundation on the bioregion's environmental conditions is the role geology plays in establishing the soil, because the rock layer determines which variety of soil type is created. The Black Forest and Vosges Mountains, being predominantly sandstone, granite, and gneiss, have highly siliceous soils, which are coarse, sandy, and often acidic, with poor nutrient and water retention. The calcium-based, alkaline limestone mountains of the Jura in Switzerland are equally poor for nutrient or water retention, as well as being highly susceptible to water erosion. Much sedimentary material initially deposited on the heights eventually arrived in the Upper Rhine Valley due to further erosion by wind and water.[9]

The Upper Rhine graben is anchored with sediments from the Tertiary and Quaternary periods, often up to depths of several hundreds of metres. Fine loess wafted in upon winter winds, while gravel, sand, flint, silt, and marl were carried onto the graben floor by rivers; under melting conditions at the end of the last ice age, plateaus of gravel, flint, and sand were formed which built contrasts to the drifts and layers of fine loess and older sediment. The abundant blanket of loess led to predominantly nutrient-rich brown soils of loamy sand and sandy loam, interspersed with equally nutrient-rich alluvial soils of sandy loam, silty loam, and clayey loam. In agricultural terms, loam retains nutrients well, accepts and retains water easily while allowing excess to drain away, and feels soft and crumbly when held in the hand. The floor of the graben was renowned for its fertility, and few disagreed with the anonymous Revolutionary of the Upper Rhine when he proclaimed Alsace to be the fairest of all places and filled with beautiful fruit, good wine, and corn, meat, and fish.[10]

Rhine and Ill Rivers

Central to the identity of the bioregion as well as its ecology, the Rhine River flows from south to north through Alsace and features prominently in its economic history as well as its environmental history. With a length of 1,232 km (766 miles), the Rhine is the only European river which has its head-waters in the Alps and flows to the North Sea; its watersheds drain 189,700 km^2 of central European soils. Flowing through several bioregions, each of the ten sections has its own name: the Anterior Rhine (*der Vorderrhein*) and the

Posterior Rhine (*der Hinterrhein*) come together to form the Alpine Rhine (*der Alpenrhein*), which bubbles along energetically for 102 km until reaching Lake Constance (*der Bodensee*). The lake, 63 km long, constitutes its own section of the river. Exiting west from Lake Constance, the High Rhine (*der Hochrhein*) receives water from five major tributaries; these include the Aare, which discharges the largest water basin in Switzerland. The High Rhine tumbles over the Rhine Falls (*der Rheinfall*) near Schaffhausen, makes a 90-degree turn to the north at the Rhine Knee in Basel (*das Rheinknie*) and, today, becomes the Upper Rhine (*der Oberrhein*) at that elbow. Medieval sources, however, often located the start of the Upper Rhine above the Rhine Knee: for them, the Upper Rhine began at the Hauenstein outcrop above Laufenburg, 39 km upriver of Basel and almost halfway to Schaffhausen.

Flowing north from Basel, the Upper Rhine flows closer to the Black Forest than the Vosges Mountains. Prior to the determination of nineteenth-century engineers to render the river navigable for commercial vessels, the Rhine was characterized, according to Marc Cioc, as possessing

> ...sinewy curves, oxbows, braids, and thousands of islands. It had a quirky, unpredictable flow, and underwater cliffs so dangerous that it spawned legends of a siren. It had sleepy fishing villages and oak-elm meadowlands on its banks. It was the site of one of Europe's most spectacular salmon runs at the Laufenburg rapids in the Swiss Alps. It contained an overabundance of allis shad (a herring-like species known colloquially as mayfish) and a modest number of sturgeon. It also supported vibrant populations of beaver, otter, bats, and birds.[11]

Management of the river by Alsatian authorities began as early as 1449, when the *Straßburg Verordnungen* were established to regulate the problems of pollution and over-fishing. Although this nineteenth-century map of the Rhine near Strasbourg prior to canalization (see Map 1.2) features a landscape which has already obviously been changed by human activity, it may serve to point towards the complexity of the river in the sixteenth century. It is extremely unlikely the riverbeds were in the same location, as the very low-lying riverbank permitted the river to change course in response to conditions such as flood, high water, or drought. The width of the medieval waterway fluctuated dramatically, sometimes up to 4 km in stretches of braiding and islets created by the periglacial conditions of the last ice age.[12]

Medieval builders slowly surmounted the challenge of erecting a bridge across the Upper Rhine. In 1225, an anchored bridge was built in Basel, and then, in 1275, another at Breisach, which, at about a third of the way from Basel to Strasbourg, was well-located for traffic between Colmar and Freibourg-im-Breisgau. The Strasbourgeois laid a *Schiffbrücke*, a pontoon bridge, across the river in 1333, and built the *Rheinbrücke*, an anchored bridge, at the same location in 1388. One-third of the latter was destroyed

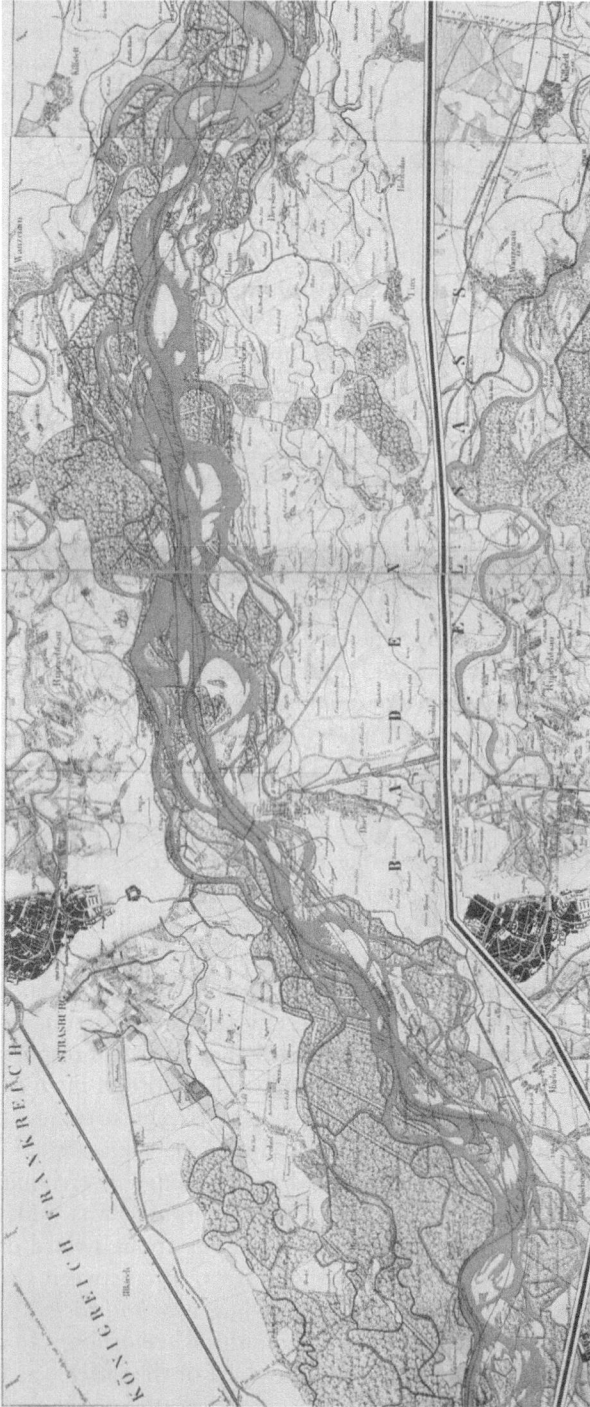

Map 1.2 Upper Rhine River prior to canalization.

by a flood in 1404 and entirely swept away in 1566. The bridge at Strasbourg was the northernmost point available to cross the river in the sixteenth century; controlling the bridge generated income for the city, and made it a centre of trade between eastern and western Europe.[13]

Headwaters for the Rhine River are largely the result of Alpine snow and glacial melt, as well as rainwater, but all the Rhine's source waters moderate each other to produce a fairly stable flow after the Rhine Knee. The Alpine Rhine and the High Rhine show a pronounced winter low and spring high, but this is balanced by Lake Constance, and the tributaries flowing into the Upper and Middle Rhine produce maximum volumes at an earlier time of year, due to being at lower elevations. High water, therefore, in the Upper Rhine under normal conditions tends to come during the summer: June and July. Winter temperatures along the river, up to Basel and the start of the High Rhine, are usually mild and the Upper Rhine freezes only in exceptionally cold winters.

Although it is generally accepted as situated beside the Rhine, it is more accurate to say that Strasbourg was built on an island in the Ill River, immediately southwest of that tributary's junction with the Rhine. Due to the notoriously shifting nature of the Rhine riverbed, it is difficult to be precise about the spatial relationship between the two rivers and the city; the distance between the cathedral and the Rhine is likely to have been much less than the 4.7 km it is today. The Ill was easily bridged and eleven of the twenty bridges used today in Strasbourg were built by 1500. The smaller river's source is found in the Jura Mountains, near the village of Winkel, Switzerland, located some 35 km south-east of Basel. The water flows quickly downhill and calmly travels most of its 216.7 km through the flat bottom of the graben on a course parallel to the Rhine. Mulhouse and Colmar, two of Alsace's largest secondary cities, are also located on the banks of the Ill. Along with the other rivers and large streams of the area, the Ill River facilitated the movement of wine and other trade goods, provided energy for water mills, washed away waste, and provided drinking water, among other things. The waterways of Alsace served a vital purpose for late medieval society.[14]

Forests, vineyards, and agriculture

Trees provided primary source material for energy needs (firewood) and stickwood production (vineyards), as well as tanning and building in late medieval Alsace. From a forestry perspective, the Upper Rhine graben is divided into three zones: plains (altitudes from 150 m to 250 m above sea level (asl), foothills (from 250 m to 450 m asl), and mountains (from 450 m to 1450 m asl). The beech and silver fir forests of the Vosges mountains, eagerly exploited by sixteenth-century miners, are known as high forests, while low or coppicing forests tended to occupy the foothills above the vineyards first introduced to the bioregion by the Romans.[15]

Since their introduction centuries ago until the mechanization of the 1950s, it was a tradition in Alsace to stabilize every single vine with a stick, using six to seven thousand pieces of wood per hectare. In the Black Forest, young firs were used for the same purpose. The regular demand by vintners added to the reasons for locating coppicing forests by settlements and vineyards.

Extending from the foothills of the Vosges onto the plain in western Alsace, grapes for making wine were the foundation of the area's late medieval economy and a prominent part of the region's agricultural activity; Rhenish wine was traded as far north as the English court, for example. On the east side of the Rhine, though, while vineyards were also common, viticulture was limited primarily to the foothills of the Black Forest. Gentler slopes and plateaus inside the forest favoured shifting cultivation focused on cereals for self-sufficiency. Staples of the peasant diet, cereals were also grown on the plain, along with vegetables and fruit. Domestic animals grazed on the graben more frequently than in the forests, which were largely the domain of wild animals. Where the lowlands were forested, oak and hornbeam featured strongly in a mixed deciduous forest. Deforestation was extensive outside of carefully managed coppice forests throughout the Upper Rhine graben, however, resulting in an underlying vulnerability to extreme flooding and earthslips.[16]

Climate reconstruction

The Spörer Minimum was the second coldest trough of the Little Ice Age, which was a volatile period of European climate approximately five centuries long and characterized by irregular temperature averages well below mid-twentieth-century norms. As mentioned in the Introduction, the relationship between climate and weather is an intimate one because they are based on the same phenomena interpreted at different scales. Weather is the atmospheric condition experienced on a daily basis. Climate is a measure of the average pattern of variation in temperature, precipitation, wind, atmospheric pressure, and other meteorological variables in a given region over long periods of time; it takes 30 years of weather observation and measurement before confident statements may be made about climate. Its study, the science of climatology, allows for comparisons across regions distant from each other in time and space. Much in the same way that individual financial transactions are the essential building blocks of that abstract domain, we know as 'the economy,' individual weather events are the essential building blocks of climate.

Weather during the Spörer Minimum attracts attention because the weather conditions of that period were frequently outside of norms desirable for Alsatian society. Other elements of the natural environment, such as the geology, rivers, soils, or forests, were fairly stable except when influenced by extreme weather events, such as heat waves contributing to forest fires or excessive rain causing floods. The volatility of the weather made it a prominent source of turbulence in an otherwise stable natural environment and a key driver of changes in the natural sphere of causation in Alsace at that time. While much of the natural world was accessible to resource exploitation by sixteenth-century Alsatians (for example, mountains were mined and soils were farmed), weather was outside of all human attempts to control or influence it, despite many ritual efforts to do so. The interactive model of socio-economic metabolism suggests that changes in the natural sphere inspire, via humanity's biophysical experience of them, changes in cultural representations of nature; attention to the key driver of change in the Alsatian natural environment is therefore merited when describing that environment.

Historical climatology, or the study and reconstruction of climatic conditions of the past, has emerged as a science during the last 50 years primarily in response to an urgent need for a basis of comparison with current climate behaviour. Research in the sub-field draws upon both the natural world and human culture for data; with respect to the former, direct measurement and proxy data provide evidence, while the principal source material for evidence from human culture is documentary. Regular weather measurements are largely unavailable before the seventeenth and eighteenth centuries, depending on region, and historical climatologists rely on proxy data from documents as well as from the natural environment for climate reconstructions prior to the onset of instrumental measurement. Details about organic phenomena (proxy data) that are sought in the documentary record include the start dates of vine or grain harvests, yield volumes, sugar content in wine, or plant phenology (that is, budding, flowering, or fruiting). Non-organic phenomena which may have been noted by authors include water levels, snow fall, the freezing or melting of bodies of water, the duration of snow cover, etc.[17]

The two data sources (natural world and human culture) provide different types and levels of information, and the integration of both types of data poses an interdisciplinary challenge, particularly when faced with criteria for qualitatively heterogeneous evidence (that is, evidence which can be reliably correlated and compared). Information from the natural environment, for example, may provide data sets which are lengthy in time, but at a very low resolution. Documentary sources, on the other hand, may provide information which has high resolution (daily, weekly, monthly, seasonal, or annual), but many are limited to the lifespan or interests of a single observer and, therefore, brief with respect to the number of years covered. Institutional commitments to recording regular and recurring events are often ideal sources for historical climatologists, but they are rare. With respect to documentary data, the primary challenge is qualitative: can the documents be trusted? Information crucial for a reconstruction of past climate may be entirely absent, false, inaccurate, or provided in an obscure manner. The historian must negotiate the perils of the Gregorian calendar reform, accepted at various years in different parts of Europe, and the hazards of regional variations on New Year's Day.

As Mauelshagen writes, errors and inaccuracies often remain. As much as possible, two or more independent witnesses are preferred to confirm weather events.[18]

Establishing the suitability of documentary sources for reconstructing the climate of a region is a rigorous process, and unfortunately, documents from Strasbourg and Alsace in the late fifteenth and early sixteenth centuries fail the test. Originality of the material under assessment is central to determining its reliability, and although there are many edited documents available from Strasbourg between 1509 and 1541 (introduced in the next section), I found only a single original source which was created by an author who could have been an eyewitness to a few of the weather events in Strasbourg and the surrounding region: Sebastian Franck's *Chronica, Zeytbuoch and Geschychtbibel* (Franck may have lived in Strasbourg from 1529 to 1531 or 1532). This surprising absence may be partially attributed to the German destruction of Strasbourg's municipal library in 1870, an unfortunate consequence of the Siege of Strasbourg during the Franco-Prussian War (1870–71); the building housed a renowned collection of medieval manuscripts (including chronicles and annals), Renaissance volumes, and Reformation archives, among other items.[19]

Although a rigorous climate reconstruction for Strasbourg and Alsace is not possible at this time, a viable substitution is available: in 2011, Oliver Wetter and Christian Pfister published a 517-year spring-summer annual temperature reconstruction based on 25 partial series of the starting dates for harvests of winter rye. The focus of the study was on Basel and its surroundings, with information gathered from the Swiss Plateau, the northern foothills of the Jura mountains, the Upper Rhine Valley, and the Swabian Alps in southwestern Germany from 1454 to 1970. Their data is partly based on harvest-related bookkeeping by institutions such as hospitals or municipalities, and partly upon early phenological observations. Evidence was evaluated and verified, with extremely early or late harvest dates corroborated with other narrative documentary reports. Then the data was sorted into four types (wage payment dates, tithe auction data, historic phenological data, and phenological network observations), with an extensive discussion about the nature and viability of each type. For the period from 1454 to 1705, the data was mostly drawn from the books of expenditures kept by the Basel hospital in which wages paid to harvest workers were recorded on a daily basis. After correcting the evidence for dating style, data type, altitude, and, in the case of tithe auction dates, estimating the start of the harvest after the tithe auction date, the 25 series were merged into a single long series. Finally, this single long series was calibrated with the homogeneous HISTALP temperature series for the period of 1774 to 1824 and verified with the same for the period from 1920 to 1970. For the purposes of their research, the scientists then compared their results with temperature reconstructions from Switzerland, Germany, Austria, France,

Hungary, and Czech Republic, as well as fluctuations in lengths of the Lower Grindelwald Glacier in the Bernese Alps after 1500 and the Aletsch glacier, whose fluctuations are primarily determined by air temperature and secondarily by precipitation. Supporting Wetter and Pfister's findings, there were significant correlations among neighbouring regions and diminished correlations with increasing distance from Switzerland, while fluctuations in glacial length were better reflected in the newly produced Basel winter grain harvest date series.[20]

Of the information produced by Wetter and Pfister, the most useful for my research is that found in a table reconstructing average temperature anomalies from March to July 1454 to 1970 (see Figure 1.1). The table includes an 11-year moving average of reconstructed temperature anomalies, with confidence bounds drawn on a ±2 × Sigma threshold. From 1473 to 1540, there are clear indications of extreme weather events during the spring and early summer in the region, particularly in 1473, 1483, and 1511.[21]

As demonstrated by the comparisons made by Wetter and Pfister, regional weather behaviour, no matter how brief a specific event may be, does not occur in isolation from neighbouring regions. This renders a brief mention of global and continental contexts advisable here. Prior to the advent of anthropogenic climate change, changes in the terrestrial climate were linked to the volume of radiation arriving on earth from the sun; that volume is affected by sunspot activity. Low sunspot activity indicates less radiation arriving on Earth, which means a cooler climate, and the reverse for high sunspot activity. From a climatological perspective, then, the context of late medieval/early modern weather events occurred in diminution of solar activity from 1460 to 1550 known as the Spörer Minimum, itself part of a lengthier

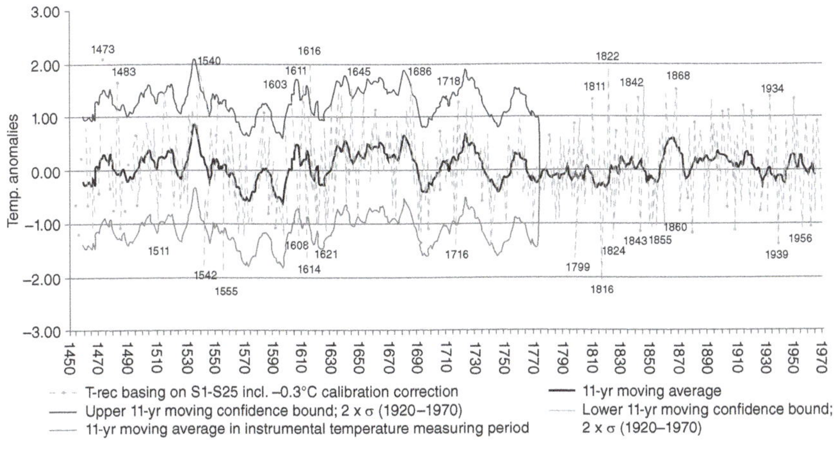

Figure 1.1 Mean March–July temperature anomalies reconstruction based on Basel WGHD series 1454–1970.

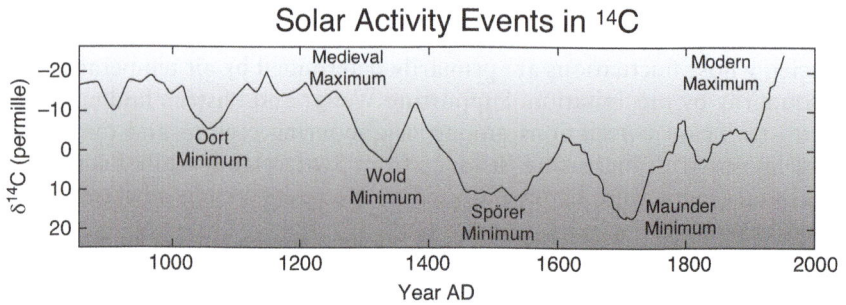

Figure 1.2 Solar activity events recorded in radiocarbon.

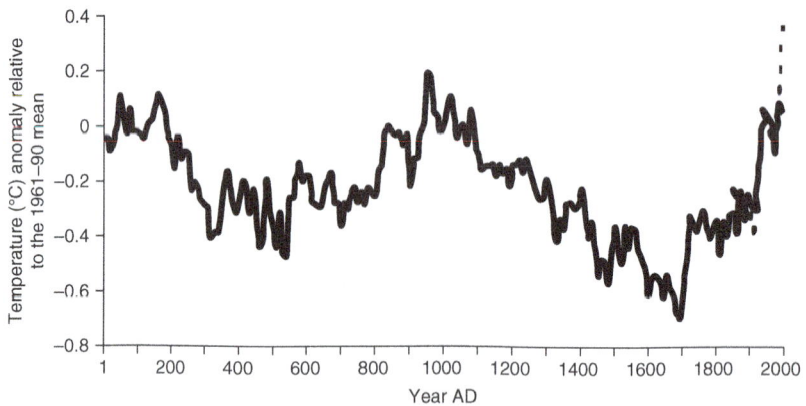

Figure 1.3 Estimations of extra-tropical Northern Hemisphere (90–30°N) decadal mean temperature variations (black line) AD 1–1999 relative to the
1961–1990 mean instrumental temperature.

half-millennium cycle of cooling average temperatures known as the Little
Ice Age (see Figure 1.2).

It must be noted that the period was not one of unrelenting cold weather;
rather, it was characterized by a series of rapid temperature fluctuations,
weather variations, and extreme weather events, with the most pronounced
phase from 1550 to 1700 (the Maunder Minimum). Moreover, the impact
of climate cooling was different in the Mediterranean basin than in northern
Europe, as befitting different (although adjacent) regional climates. Despite
this, European temperatures show an average decrease of 1°C during the LIA
as compared to late twentieth-century average temperatures and, in regions
of central Europe, as much as 2°C below average during the Maunder Minimum, the Little Ice Age's coldest period (see Figure 1.3).

The period of my research takes place during the Spörer Minimum, the second lowest trough of sunspot activity and second coldest average terrestrial

temperatures during the most recent millennium. Extreme weather events were and are not unusual for Alsace, whose climate is characterized as a mixture of maritime and continental, or semi-continental. The frequency of such extreme weather events, however, is contextualized by the wider climatic context of Europe, the northern hemisphere, and the planet. Moreover, the relationship between weather and climate is not direct: as accustomed weather patterns change under the influence of changes in atmospheric and marine temperatures, a warming climate may still produce extremely cold temperatures and a cooling climate, as was under way during the last half of the fifteenth century and early part of the sixteenth century, may also produce heat waves such as that of 1540. For more precise information about the weather of Strasbourg and Alsace, we must look to narrative evidence found in documentary sources.

Alsatian weather 1473–1540: sources for the knowledge thereof

As noted by Stefan Brönnimann, "… it is often not the average conditions over a month, season, or year that matter for society, but weather events, which are not resolved in climate reconstructions."[22] For the pre-instrumental period, documentary evidence is key in determining the timing and severity of weather events and there are many edited volumes reproducing manuscript originals which contain information about the weather in Alsace during my research period. Assembling them together permits me to offer an approximate annual report of the weather events from 1473 to 1509 (Chapter 2), 1509 to 1525 (Chapter 3), and 1525 to 1540 (Chapter 4) in and around Strasbourg and Alsace.[23]

Of these edited volumes, the most relevant are those which were published in the city and take events there as their focus. They are Johann Jakob Meyer's *La Chronique Strasbourg*, Father Martin Stauffenberger's *Annales der Barfuesser zu Strassburg, de anno 1507 biss 1510*, *Les Éphémérides de Jacques de Gottesheim*, and *Les Collectanées de Daniel Specklin*. 'Fortsetzungen des Königshoven. Straßburger Zusätze' and 'Straßburger Jahrgeschichten (1424–1593)' are included in Franz-Josef Mone's *Quellensammlung der badischen Landesgeschichte*. As well, two pertinent Alsatian chronicles published before 1870 will be included: P.F. Malachias Tschamser's *Annales oder Jahrs-Geschichten der Baarfüseren oder Minderen Brüdern S. Franc* and Xavier Mossmann's *Chronique des dominicains de Guebwiller*, the first from a Franciscan monastery at Thann and the second from the Dominican convent at Guebwiller. A discussion of these sources follows here. Single entries from other sources will be included in my weather report, but will not receive detailed attention as they are not central to it.[24]

Authorship of *La Chronique Strasbourg* is attributed to Johann Jakob Meyer, a mysterious writer about whom not even birth and death dates are known; the chronicle at the core of the document is believed to have been first published in 1587. Relying heavily on Jakob Twinger von Könighoven's

Chronik until the early fifteenth century, *La Chronique Strasbourg* gives a history of the city from its origins until 1711. It is divided into chronologically ordered chapters, one each devoted to events concerning cloisters and other religious establishments, the dioceses, particular buildings, regulations, bishops, *ammeisters*, agreements between the Emperor and the City, military campaigns, catastrophes, and fires. Unfortunately, the original manuscript was destroyed with the Strasbourg archive in 1870; fortunately, shortly before the fire, Rodolphe Reuss had copied much of it for the research project which resulted in his 1873 volume. Material which Reuss took directly from Twinger was only cited, not reproduced verbatim, and the following paragraphs are based on his notes and conclusions.

Primary among these conclusions was the observation that the 1587 manuscript from which Reuss made his transcript was unlikely to have been the original. Instead, he suggests that it was a copy made for unknown reasons by Philippe Engler, secretary to Strasbourg's council of XV. Whether or not Reuss' suggestion is accurate, identifying the original author of *La Chronique Strasbourgeoise* is problematic due to the editor's disclosure that four or five distinct hands were visible in the original manuscript, dispersed among the chapters and occasionally on pages added after publication. The first handwriting stopped at a different date in each chapter; a second hand was added to the first, the earliest date of which was 1511 and the latest 1611. A third hand added a few items from 1624 to 1628. A fourth hand started adding items from 1619, and apparently started to use the chronicle as a journal and records events until 1711. Reuss suspected the presence of a fifth hand in two or three isolated items recorded from years 1634 to 1642.[25]

There is no way to determine if the author of the earliest handwriting in *La Chronique Strasbourg* was a witness to the weather events included in the manuscript nor to determine if all events from the period were included. With the manuscript's destruction, future exploration of it is impossible. Nevertheless, the chronicle covers events which occurred during the entire period under study and, therefore, provides information which merits its inclusion in my weather report.

The next two selections were included in a booklet donated to the Strasbourg municipal library by the heirs of Professor Frédéric Reussner, who received it from his step-father, the former curator of the library M. André Jung (d. 1863). Both appear to be a partial copy of an original; the rest of the booklet contains material from the end of the sixteenth and beginning of the seventeenth centuries. Alsatian historian Rodolphe Reuss edited both selections into articles published in *Bulletin de la Société pour la Conservation des Monuments Historiques*. Of the first, the *Annales der Barfuesser zu Strassburg, de anno 1507 biss 1510* written by friar Martin Stauffenberger, only eight pages exist. Little is known of Stauffenberger other than what is available in the *Annales*, where he wrote that he was a novice in the Franciscan convent in Strasbourg in 1507 and became the monastery's Receiver on Saint Sebastian's Day (*20 Jan*) in 1510. He held this role until New Year's Day 1511.

Reuss offers a simple transcription of the pages, without commentary or edits; they cover events from 1507 to 1510, with the inclusion of an entry for 1482. With the only available edition a nineteenth-century transcription of a late sixteenth-century copy, there is no way of assessing the relationship of the surviving material to actual weather events. It may be an eyewitness account, but it is unlikely that confirmation will be achieved and the document provides only partial coverage of the period of research. Despite this, the author's role in Strasbourg's Franciscan monastery promises a sufficiently reliable witness to weather events from 1507 to 1510 for the inclusion of entries in my weather report.[26]

Another eight pages of Reussner's booklet held *Les Éphémérides de Jacques de Gottesheim*, which covered the years 1524 to 1543. The similarity in the number of pages with the previous source deepens suspicion that both *Les Éphémérides* and the *Annales* were partial copies of originals now lost. *Les Éphémérides*, though, offers clues that the copyist of both manuscripts is likely to have been a member of Strasbourg's *Rat*, as Gottesheim's papers were part of the official records of that body's secret councils and few others held access to those archives in the sixteenth century. Gottesheim was a doctor of law who received the Franciscan orders; he was made prebendary of the Great Choir in 1517, as attested by an annotation from the anonymous copyist. Gottesheim remained a loyal Catholic throughout the religious conflict which followed. By 1522, he held the position of episcopal vicar and, as might be expected from one occupying that position, was requested in December of that year to develop the case against Matthäus Zell for a charge of heresy (the case was developed, but not pursued). As the city began to follow the Reform, though, Gottesheim did not join the Catholic exodus but, rather, obeyed a municipal order to either become a citizen or leave the city by accepting citizenship on 1 February 1525. Becoming a citizen of Strasbourg did not change his sympathies and he was accused of espionage more than once. Despite this, he remained on good terms with the city's elite and boasted of dining with Bucer, Zell, Capito, and Sturm; Katarina Zell invited him to her house for a good theological debate. He continued to enjoy the approval of William von Hohnstein and, upon the latter's death, the support of the new Bishop of Strasbourg, Erasmus Schenk von Limbourg. The last date for which Gottesheim is known to have been alive is 1543, and it may be assumed his death happened either later that year or shortly after. The anonymous copyist was hostile to his source, however, and the version of *Les Éphémérides* which survived may not be trusted for fidelity to the original. As well, the eight pages of material only cover from 1524 to 1543, the last few decades of the weather report. However, as in the case of Stauffenberger, Gottesheim's proximity to events in Strasbourg offer enough promise of an eyewitness to weather events that entries from *Les Éphémérides* are included in this weather report.[27]

Daniel Specklin (1536–1589) was born and died in Strasbourg and there has been much scholarly interest in him. As he was unlikely to have remembered and recorded weather events at four years of age and younger, though, only a summary of his life and career will be provided here. Apprenticed in Strasbourg as both an embroiderer and a woodcarver, by 1555, he was in Vienna and toured Anvers, northern Germany, Denmark, and possibly Sweden in 1561. By the end of that year he was in Austria, where he met Hermann Schallantzer, entered his service and became an engineer and architect. This proved to become the foundation of his career, as he was involved in building the fortifications of Ingolstadt in 1576. The next year, he was named city architect of Strasbourg, a post created especially for him. He was popular throughout Alsace and contributed to the fortifications of Ensisheim, Basel, Lichtenberg, and Heilbronn, among others. Although written on behalf of Strasbourg's *Rat*, along with two other books (*Architectura von Vestungen* and a lost volume on the improvement of the city's fortifications), requests for edits and amendments delayed the completion of his chronicle. Unfortunately, Specklin fell ill and died before developing his collection of extracts and personal memoires beyond a chronologically ordered list.[28]

The disorder of the initial collection is reflected in *Les Collectanées de Daniel Specklin, Chronique Strasbourgeoise du seizième siècle*, edited by Rodolphe Reuss. The original was lost in the fire of 1870, leading to a community endeavour from which emerged a compilation of eight different manuscript particles. The haphazard manner in which the material of *Les Collectanées* arrived into the editorial control of Reuss makes it variably reliable for the purposes of this weather report. Moreover, other than Königshoven's *Chronik*, it is impossible to give an account of the sources Specklin used for events he did not personally witness, although the dates his life spanned suggest they may have included interviews with eyewitnesses. Specklin's status and role in Strasbourg civil society combine with his relative proximity to events, however, to allow the inclusion of entries from *Les Collectanées* in the weather report.

The authors of 'Fortsetzungen des Königshoven. Straßburger Zusätze' and 'Straßburger Jahrgeschichten (1424–1593)' are unknown, as is unfortunately also true for *Annales oder Jahrs-Geschichten der Baarfüseren oder Minderen Brüdern S. Franc.* and *Chronique des dominicains de Guebwiller*. All that is available is a few words about Franz-Joseph Mone (1796–1871), the editor of the volume wherein they are published, *Quellensammlung der badischen Landesgeschichte*. Born in Mingolsheim bei Bruchsal, a town north of Karlsruhe, itself north of Strasbourg along the Rhine, Mone held a position as historian at the University of Heidelberg and on the board of the institution's library. He was known for his work with Old German literature and language, as well as the *Zeitschrift für die Geschichte des Oberrheins* and the collection *Quellensammlung der badischen Landesgeschichte*. Following his career there, he moved to Karlsruhe, where he lived as a private scholar, edited *Karlsruher Zeitung*, and served as the archivist and director of the Baden Generallandesarchivs until his retirement in 1868. Although little is known about the authors or the sources he presents in *Quellensammlung der badischen Landesgeschichte*, their focus on Strasbourg and surroundings allows relevant entries to be included here.[29]

François-Antoine Tschamser (rel: Malachias; 1678–1742), editor of *Annales oder Jahrs Geschichten der Baarfüseren oder Minderen Brüdern S. Franc. Ord. inßgemein Conventualen genannt, zu Thann*, became the Father-Guardian of the Franciscan monastery in Thann, on the other side of the Grand Ballon from Guebwiller, and served in that capacity from 1722 to 1739. During that time, he gained responsibility for ordering the Franciscan convent in Haguenau as well. Using unidentified chronicles from Franciscan archives and, for recent events, oral accounts from his friars and visitors, Tschamser assembled a chronicle of events in Thann and the wider region of Alsace from 1182 to 1700; he first published the four volumes of his work in 1724. The edition used here is the third re-publication and the earliest available. Unfortunately, he also included items from other regions or the same facts under more than one date. Many of the regional weather events he cited, though, are regularly independently corroborated by other sources and the entire period of research is covered by events in the *Annales*. These considerations combine to allow for sufficient confidence to allow Tschamser's entries to be included as evidence in my weather report.

Xavier Mossmann (1821–1893), editor of *Chronique des dominicains de Guebwiller*, identifies his source as a chronicle found in the library at Colmar, where it was part of a collection inherited from the library of the Imperial Abbey at Murbach, a noted Benedictine monastery in the southern Vosges mountains between Colmar and Mulhouse. The last author of the chronicle is reported as prior Séraphin Dietler, whose final entry was in 1723; Dietler also edited the chronicle into the form adopted by Mossman. Like many other sources used here, the text is doubly removed from events and segments may have been copied from the chronicle of Thann. Nevertheless, entries may be used to corroborate other accounts and will be included to provide indications of the scope and nature of regional weather events. Sources for the *Chronique* are likely to have included Hans Stolz's *Ursprung und Anfang der Statt Gebweyler*; little is known about Stolz, but his personal account of weather and other events from 1502 to 1540 is an important documentary source for the weather events and some of their social consequences near Guebwiller in southern Alsace.[30]

These are the primary sources I used to build my weather report in Chapters 2–4, along with fragments from Tambora, the climate and environmental history collaborative research environment online database. The database is a large collection of original text quotations concerning weather and other environmental events, along with bibliographic references. Together with the sources provided above, a reasonable report on Alsatian weather between 1473 and 1540 is made possible. It is important to note that while historical climatological research continues to advance, it is impossible to ensure that this weather report for those decades is exhaustive. Significant weather events may have been omitted from the accessible records, or are perhaps held in an archive I have not yet explored. Five hundred years have passed since these events and it is a gift that knowledge of the weather in a single bioregion of the Holy Roman Empire has endured through the centuries. Despite this caution, though, enough information has survived that a reasonable weather framework for related Alsatian behaviour can be established.

An outline of Strasbourg and Alsatian society at the turn of the sixteenth century

At the dawn of the sixteenth century in the common era, an estimated 20,000 souls were living in Strasbourg; the population of Schlettstadt (*Sélestat*), the second-largest town of the Upper Rhine Valley, is likely to have reached 4000. Other than Strasbourg and Sélestat, the population of other towns in the Upper Rhine Valley would not have reached more than 1000 people. In 1500, the population of the area covering today's German state is estimated to have been 20 million people, with the highest population density in southern Germany (Alsace, Swabia, and Bavaria), on a continent whose total population is gauged as having been around 70 million people. Disease and illness, including the bubonic plague, killed between 10% and 20% of town- and city-dwellers per generation, interrupting a regular increase of people within city walls. Typical cities of sixteenth-century Europe are today recognized as having been 'population sinks' – centres of disease with a high death rate, replenished not by reproduction but through immigration.[31]

A limitation: this general information about population and population density cannot identify the exact composition of specific audiences in Strasbourg's *Liebfrauenmünster*, in a pasture beside Bernhardswiller, or in the lecture halls of the *Schola Argentoratensis* and, therefore, precisely who may have listened to sermons delivered by Johannes Geiler von Kaysersberg, Clemens Zyegler, or Wolfgang Capito. Even when illustrious personalities are known to have listened to Geiler's sermons, such as Holy Roman Emperor Maximilian I and his second wife, Bianca Maria, they were two of a possible 3000 listeners in the nave on that day in August 1504.[32]

Nevertheless, the cathedral was and remained a centre of activity in Strasbourg, open and accessible to all the city's inhabitants and visitors. Documents such as the *Book of Donors* (a record of donors and donations to the *Liebfrauenmünster* from 1320 to 1521) show that the Cathedral was beloved by many community members regardless of their status or station, and available demographic information allows indications towards the composition of an average audience. What follows is a brief description of Geiler's society following a model articulated by the preacher himself in 1509, when he told his listeners that Aristotle divided all the cities and the people in the cities into three parts: the nobility, the citizens (who enjoyed full political rights and the responsibility to provide defenders for the city if needed), and the artisans (working people such as tailors or shoemakers). His deviation from the traditional threefold medieval conception of the social order (*orantes, laborantes,* and *bellantes*, or, as known in the local dialect, *praelaten/praelatenstand*, *landschaft*, and *ritterschaft* and, in English, preachers, workers, and fighters) provides a neat framework for this description of society in Strasbourg and Alsace around 1500, any member of which may have entered the Cathedral to listen to one of Geiler's sermons. Changes from this model will be noted in future chapters as they occur.[33]

The right to bear arms, which was the general defining feature of a 'gentleman' in late medieval Europe, was materially based upon the ability to live from one's investments, either in land or trade, rather than by labour. Once this criterion was established, however, further access to status or power could depend upon various local or historical factors. Regarding Strasbourg, a charter granting the status of *Freie Reichsstadt* (Free Imperial City) had been awarded to the municipality in 1262 by King Philip of Swabia, thereby establishing its independence from the Bishop of Strasbourg after the battle of Oberhausbergen. A key aspect of this charter was the direct submission of the city to the Holy Roman Emperor personally; no intervening magnate, Prince, or Bishop (or Prince-Bishop) could exercise authority over Strasbourg in the Emperor's name. The municipality gained the liberty to conduct its own military and economic affairs and, if desired, to send a representative to the Imperial Diet. Ten smaller urban centres in Alsace achieved a similar status: known as the *Zehnstädtebund* or Décapole, the towns of Colmar, Haguenau, Kaysersberg, Landau (from 1521), Mulhouse (until 1515), Munster, Obernai, Rosheim, Sélestat, Turckheim, and Wissembourg joined in a *Städtebünde* (a military and economic alliance between cities) in 1354. While successfully protecting their political and economic independence, members of the alliance did not grow to equal Strasbourg's population and influence, largely due to the latter's construction of and protection for the northernmost bridge across the Rhine River at that time; controlling the bridge generated income for the city and made it a centre of trade between eastern and western Europe, as well as a military target. The independence enjoyed by citizens of Strasbourg and the *Zehnstädtebund* towns, however, was not shared by Alsatians in general; authority over rural people was exercised through a confusing network of land ownership and military capacity held in a complex mixture of clerical and secular hands.[34]

The most influential among the clerical rural landowners of Alsace – certainly able to equip and provide mounted knights when required – was the Bishop of Strasbourg, ousted, as mentioned above, from his cathedral seat in 1262. As landgrave of the region, a position roughly equivalent to that of an English count, the office's occupant was responsible for a variety of regional affairs, including safe-conduct for travellers and the legitimization of bastards. He also held ultimate responsibility for judgments made at ecclesiastical courts in Alsace, although courtroom duties as a judge were frequently delegated. He and the Bishop of Basel exercised civil authority that was delegated from the Holy Roman Emperor, as did the abbotts of monasteries at Murbach, Münster, Gengenbach, Schuttern, Saint Blasien, and Saint Peter in the Black

Forest. The authority of the Bishops of Metz and Toul derived from the Duke of Lorraine, a factor which would prove pivotal in 1525. On the secular side, where military capacity was higher, noble rural landowners in the Upper Rhine Valley included the Count Palatine (the Margrave of Baden-Baden), the aforementioned Duke of Lorraine, the imperial house of Austria, the Duke of Wurtemberg, three margraves (similar to dukes) from the Basel area, the Count of Hanau-Lichtenberg, the Count of Bitche-Lichtenberg, the Lord of Blâmont and the Baron of Morimont.[35]

Where the governance of the Free Imperial City of Strasbourg was concerned, along with the direct appointment of individuals, societies known as *constefeln* were a primary method of mobilizing political authority. The term *constofeln* originally identified the right and means to stable horses within city walls; by the late fifteenth century, it referred to district-based societies of the city's elite. Characteristically composed of guildsmen grown rich after decades of commerce, episcopal functionaries, and the descendants of noblemen remaining after the mass eviction of their peers in 1419, *constofeln* members usually held common cause with rural landlords because, along with the city corporation itself, they were frequently landowners themselves. Although the city was independent of any feudal overlords except the Emperor himself, there were no legal barriers preventing Strasbourgeois from owning land in the region or further afield, and exercising commensurate authority over it. The prominent Sturm family, for example, independently of the *Freie Reichsstadt*, held fiefs of the Holy Roman Emperor, the Elector Palatine, the Bishop of Strasbourg and a number of lesser Alsatian lords; the Sturms were also seigneurs of the village of Breuschwickersheim, a few miles southwest of Strasbourg.[36]

Almost two-and-a-half centuries after the establishment of Strasbourg's charter, civil authority over the city was exercised through an executive, a senate, an assembly, and a complex network of councils and committees. The most important of these last were two privy councils: the XIII and the XV, collectively known as the XXI (understood to be the total number of XIII and XV councillors who were neither *Ammeister*, *Stettmeister*, nor senators in a given year). The Senate (*Rat*) was the oldest body and consisted of 30 senators who held office for two years (ten noblemen selected by outgoing senators and 20 guildsmen appointed by their guild). During the sixteenth century, senators acted as the high court of Strasbourg on Tuesday and Thursday mornings, while on Monday, Wednesday, and Saturday mornings, they met in joint session with the XIII and XV. This joint session formed the highest legislative body of Strasbourg: the Senate and the XXI (*Rat und XXI*).

The Senate and XXI elected the *Ammeister*, who served for one year as ruling *Ammeister* and then five years as *Altammeister* (former Ammeister); his service asked him to preside over the council of XIII, the Senate, and the Senate and XXI when they gathered; to dine in public at his own dining hall; and to stand as the head of the government when necessary. The XIII was the senior and most influential privy council, focused mainly on war and diplomacy; the ruling *Ammeister* was assisted by four of the five most recently elected *Altammeisters*, four *Stettmeisters* (nobles appointed from the *constofeln* who rotated quarterly through the office) and four guildsmen. Established in 1433, the council of XV concerned itself primarily with domestic affairs, although it could also hear legal appeals from municipal courts. It was composed of five patricians and ten guildsmen (none of which could be *Ammeister*, *Altammeister*, or a current member of the XIII).

The *Rat*'s many administrative committees oversaw much of the social and economic functioning of Strasbourg and its surrounding territories. There were boards for the poorhouse (*Almosenherren*), the fair (*Messherren*), the arsenal (*Zeugherren*), and the hospitals (*Pfleger der Blatterhus, Spitalpfleger*); for the schools (*Scholarchen* or *Schulherren*), construction (*Bauherren*), and civic property (*Allmendherren*), and for each district (*Amt*) of the civic territory (*Landherren, Landpfleger*). Committees were usually composed of three men; each privy councillor was obliged to participate on as many as five of these committees and, occasionally, some held their posts for decades. Holding civic office required an oath to the city; the annual avowal of the Senate and XXI became a municipal ceremony known as the *Schwörtag* held on the first Thursday of every January. The newly elected were required to gather on a platform in the public square outside the Cathedral's western doors and, before the people, swear to uphold the city's Constitution.[37]

Only citizens of Strasbourg could represent a guild, while that status was not required of *constefeln* appointees; citizenship required an oath of loyalty to the city and the duty to provide soldiers and arms for its defence. For an emigrant to become a citizen, they were obliged to mount an application to the *Magistrat*, which was typically granted upon marriage or upon the payment of a fee. Various citizenship categories were available: *Bürger*, those who lived in the city and enjoyed the full political and economic rights of citizenship, *Ausbürger*, who lived outside the city but remained subject to Strasbourg's laws, and *Schultheißburger*, who resided in the city and could receive alms, but who were without political rights (that is, could not vote in guild affairs nor represent a guild to the *Rat*). This last was an intermediate step towards achieving citizenship, and was often used as such by recent arrivals to Strasbourg at all levels of society.[38]

Social differences between the nobility and the merchants found further substance in the political realm of the city, where the *constofeln* societies, for example, could revise their own statutes and determine membership. Membership was by nomination, which meant that most members were personally linked to the wider, even regional, networks of land-owning wealth that provided access not only to *Stammsitz* (ancestral residence) but also to informal systems of communication and opportunity leading to civil authority. The Guilds, however, from which most merchants emerged, saw their constitutions and by-laws vulnerable to review by the council of XV, and hopeful office-seekers faced an election for the privilege of representing a guild within the city's administrative structure.[39]

Over the course of the fifteenth century, the increasing prosperity of certain guilds and particular guild families led to a situation where the representatives occupying the seats of government often were not themselves workers, but, instead, merchants at least as wealthy as the lower orders of the nobility. Such men benefited from a regressive municipal tax scheme, where only the first 16,000 *gulden* of capital in one's possession was taxed; in 1501, Geiler denounced it from the pulpit as the 'stabling-money' of the rich. The Upper Rhine was – and remains – a proverbially prosperous region, and, located as it was on axes of trade both north-south (from the Mediterranean to the Baltic) and east-west (the city controlled the northernmost bridge over the Rhine River), merchants took advantage of its location to speculate in the cattle trade. Grain and wine were also exported and, when conditions were favourable, subject to speculation.[40]

Whether wealth was gained by inheritance, appointment, land ownership, or trade, its possession meant having access to civic power in Strasbourg. Although the nobility were banned from engaging in mercantile activity, Brady notes that 'great wealth was relatively widely distributed through the corporate hierarchy [such] that no single aristocratic type could be exclusively identified with any single corporate institution.'[41] For the aspiring merchants of Strasbourg, the elevation of one's status from that of mere wealth to *constofeln* membership was elusive but not impossible. Confident social intercourse between the two groups is demonstrated by such occurrences as intermarriage, the bestowal of extinct titles upon aspirant merchants, significant investment in mercantile activities by members of the nobility, and shared service to the territorial principalities.

Municipal government, however, was not the only avenue within city walls for the exercise of noble authority. Governance of the *Liebfrauenmünster* cathedral was exercised by the cathedral chapter, control of which had been removed from the Bishop of Strasbourg with the city's charter in 1262. By 1500, the chapter consisted exclusively of noblemen whose families held positions in the Imperial Diet; they were not necessarily natives or citizens of Strasbourg. It was a site of undiluted aristocratic power within the city and region, as the chapter held property in the city and elsewhere; this was the body which held Geiler's contract. The pervasive nature of its influence in Alsace is demonstrated by circumstances in late fifteenth-century Sélestat, where the *Liebfrauenmünster* chapter owned a benefice whose funds also covered the salary of the town minister. Sélestatians' unhappiness with rapid turnover and a general incompetency in the officeholders failed to find remedy until 1503, when, after decades of complaint, the chapter agreed to only receive its revenue when the pastor actually lived in Sélestat and fulfilled his pastoral duties to satisfaction.[42]

In fairness, while materially benefiting from their association with the Cathedral, the city's elite also contributed directly towards its functioning and maintenance. Evidence for this can be found in obituary manuscripts from the Cathedral; such documents were compiled by church officials to celebrate bequests, whether for particular ecclesiastical purposes or for an institution as a whole. They provide important information about donors and their priorities. One such manuscript was created in 1508 on behalf of the Cathedral choir: identified as *ABR H 1613*; it shows that bequests were made to the choir until c. 1530. Fifteen bishops, seven dukes, one count, and more than one hundred lords donated to the choir; the predominantly male gender and preponderance of surnames from the high nobility among the 829 donors indicate a strong likelihood that most donors were clergymen or their families. These donations ensured that the choir's presence was felt in the cathedral on a regular, if not daily, basis.[43]

A more representative record of donors to the *Liebfrauenmünster* is found in the Book of Donors, which recorded offerings on the altar of the Chapel of the Blessed Virgin Mary from c.1318 to 1521. While donations in *ABR H 1613* are mostly from aristocracy and divided almost equally between religious and lay donors, 88% of those whose status was identified and who gave goods, property, or cash to the Chapel of the Blessed Virgin Mary were members of the lay community. Among them, the largest single group were tradesmen and tradeswomen; people from 86 different trades donated to the Blessed Virgin Mary, including bakers, notaries, shepherds, and a wet nurse. Extremely humble gifts, such as a single kettle or a few pence, are listed alongside items of extraordinary generosity, such as a house or 200 *Pfund-Pfenning*. It is impossible to identify individual motivations for donating beyond a desire to be involved in the material, social, and spiritual fabric of the Cathedral. However, that the humblest spared resources for the altar alongside the wealthiest speaks to the broadness of its appeal as well as the engagement of lower-status citizens in activities taking place in the nave.[44]

As well as participating alongside the elite in the life of the city's largest religious institution, the artisanal and lower classes of Strasbourg also shared with them familial ties, personal associations, or economic relationships with people in the surrounding countryside. Francis Rapp notes that, for instance, among the wealthiest rural peasants, it seems to have been the custom to let a member of the family settle in Strasbourg to act as a grain dealer. Such families were likely to be tenure-holding tenants who provided rents to a secular or ecclesiastical landlord in exchange for use of the land. Tenure such as this was regularly passed down to heirs, resulting in inter-generational stability for both tenants and landlords. In contrast, a distinctive element of serfdom was the personal bond between serf and overlord in addition to their tenurial relationship, a bond frequently caused by indebtedness. Serfs could be relocated by a lord without their tenure (from one estate to another) and their ability to conclude legally binding contracts was more limited than that of other peasants. However, serfs certainly maintained a legal personality and were able to seek court protection or defend customary rights by legal means. Serfs were not outright slaves, but often the condition was inherited by descendants.[45]

A general flow of migration to the urban centre was on-going, with bigger waves at times of crisis in the countryside and despite the potential health consequences of being in the city. Impeding this movement of people from rural Alsace to urban Strasbourg was official discouragement; punishment for the abandonment of a farm, for example, included the confiscation of all goods left behind, and one can only imagine the peer pressure to remain on one's farm created by the legal obligation of former neighbours to reimburse a lord for losses caused by someone's move to the city. If an individual was embroiled in serfdom, restrictions were even tighter. In 1512, for example, Jean-Ulrich de Ferrette demanded the people of Mulhouse return one of his serfs who had settled among them and in 1515, Emperor Maximilian I felt authorized to prohibit Upper Swabian serfs from settling in Strasbourg. [46]

The countryside was not noticeably depleted by the trickle of people to the urban centre, although a scarcity of records makes it difficult to find clarity about annual or even decadal changes in late medieval rural Alsatian demographics. Records from rural Ban-de-la-Roche (50 km south-east of Strasbourg, in the Vosges mountains), for instance, suggest a gentle growth rate by succeeding generations in the rural parts of Alsace: 73 houses and 383 inhabitants in 1489 rose to 107 houses and 560 inhabitants in 1534. Rapp disagrees, however, saying that changing demographic figures like this may reflect a concentration of the population, rather than population growth, as people migrated to larger and safer settlements in the region. As well as flight from dangerous conditions, motivation for peasants to move away from their farms may be found in the material and social conditions they endured, the difficulties of which were heightened by social inequality despite the natural wealth of the land. George Bischoff points out that a year of harvest failure, with high prices of grain followed by reports of *grosse sterbet* (great death), may be beneficial for the vendor of grain but catastrophic for the purchaser. He draws a portrait of a peasantry burdened by taxation and debt, restricted from emigrating, and infuriated by speculation and usury.[47]

Consider the example of peasants living in the Bruche Valley, in the Vosges Mountains west of Strasbourg. According to Francis Rapp's 1979 analysis, in Alsace during average conditions of the first half of the fifteenth century, rents and taxes did not consume all that was produced in years when conditions allowed for good harvests, and prices of a quarter of cereal almost invariably fluctuated between 40 and 90 *pfennig*. From 1450 onwards, resources formerly held in common were privatized at a heavy rate or development occurred without regard for communal welfare. A common complaint, for example, was the construction of a fish pond by landowners who were indifferent to downstream consequences. Increasingly severe price fluctuations arrived after 1460, with prices for a quarter of grain reaching as low as 26 *pfennig* and as high as 160 *pfennig*; similar fluctuations in wine prices coincided with those of grain. Meanwhile, the Bruche Valley peasants saw their taxes rise by 20% between 1482 and 1502, a consequence of the indebtedness of their overlord, Bishop of Strasbourg Albrecht von Pfalz-Mosbach. Although a quarter of the annual tax might be forgiven, or a little cheating might occur without penalty, the sum represented an increasingly large portion of a small annual income. Private loans could be easily found with a small piece of land as a guarantee, but failure to meet a single repayment could result in the loss of the security – and, in the sixteenth century, frequently did. Compounding this vulnerability, the Upper Rhine was an area where partible inheritance patterns were practised, which made inter-generational continuity of property ownership more susceptible to individual fortune.[48]

The Alsatian peasantry could fairly be described as having been squeezed by social contradictions. While fiscal demands upon them during the late fifteenth and early sixteenth centuries increased along with their population, war, pestilence, or weather-related harvest failure destroyed their ability to fulfil those demands. Similar to Peter Blickle's 1977 study of nearby Swabia, where 40–50% of the population were without property, Tom Scott's 1997 research in the Upper Rhine confirmed contemporary complaints of an epidemic of vagrancy and begging by the poor, the homeless, the unemployed, and discharged mercenaries. In 2010, Georges Bischoff explicitly attributed rootlessness to economic hardship and extended the list of vagrants and beggars to include a wider range of people than commonly assumed, including the respectable: a journeyman travelling from city to city, a highway robber, a touring student, a peddler, a Jew, a pilgrim from Saint-Jacques, a band of *landsknechte*, a troop of minstrels or a caravan of merchants.[49]

Unequal access to the resource base (land, livestock, or fish) meant that many were regularly hungry, poor, and ill-housed, even while economic analyses of the Upper Rhine provided the illusion of plenty for all. Harvest failure or even poor harvests led to increasing prices which created conditions of dearth or famine for the poorest strata of society, as occurred during almost half of the years between 1473 and 1509. When faced with such challenges, though, peasants were not without resources. Both in the *Weisung*, an annual ritual reaffirmation of subjection, and through direct petition to overlords, people are known to have made attempts to improve their circumstances, and some were occasionally successful.[50]

Aside from direct or indirect appeals to overlords, another resort of rural people when faced with overwhelming threat to life and limb was to physically remove oneself to urban centres. Geiler, among others, considered the welfare of the poor a priority and was involved with Strasbourg's welfare system throughout his career. As suggested by the chronicler Hieronymus Gebwiler, author of the *Straßburger Chronik* and the *Schlettstadter Chronik*, Geiler's work contributed to meeting the basic needs of a thousand mendicants over the winter of 1509/1510 – equivalent to nearly 5% of the city's population. Figures from the mid-sixteenth century (when head counts by welfare donors became more regular) show that during harsh winters, peasants from the surrounding areas cast themselves upon the compassion of the citizens of Strasbourg. For example, during the winter of 1565/1566, outside the range of this study but representative of the tradition, the extraordinary cold brought an estimated 5000 people into the city, temporarily increasing the population to over 25,000.[51]

The Strasbourg community at any moment, then, likely included extra-mural visitors along with members of the entire socio-economic spectrum from the elite to the indigent. It is not beyond the realm of probability that such individuals would have visited the cathedral, listened to the famous preacher, and participated in the life of the nave.

Later in the sermon which framed the above-mentioned tripartite division of his society, Geiler introduces the clergy. He allocates them among all three categories (workers, merchants, and nobility), suggesting that the clergy are human equivalents to the ants who stand by the column to guard and guide the other ants: preachers among the workers, priests with the merchants, and the bishops and doctors with the nobility. He elaborates on this last as follows:

> The third ant stands by the path so the others will know the way, and the third men, those are the nobles, Prelates, and particular Bishops and Doctors. Some say, what do the nobles have to do with producing, they are idle. They are not idle when they are at home reading in books to see how they should advise the Council and passing judgment and others read about how to know the right way and punish evil. That is a proper employment. The Bishop to preach and to teach about the way to heaven, etc. (Shepherds and teachers) on that account Saint Paul bound together these two ministries, shepherds and teachers. So that they should lead their little sheep as shepherds, and direct and teach them as teachers.[52]

Geiler's description of preachers as occupying a leadership role in society was hardly accidental, casual, or naive. As established by Rita Voltmer, it is clear that Geiler understood the integrated accumulation of moral, social, and political power exerted by the Roman Catholic Church in his time, and that he did not shy away from exerting it in his attempt to establish his ideal of the New Jerusalem in Strasbourg. An early audience upon whom he focused his energy was Albert of Palatinate-Mosbach, Bishop of Strasbourg, his court, and the higher clergy; to them Geiler delivered the opening address in a diocesan synod in 1482, summoning them to purify their lifestyles with a detailed platform of specific renunciations and proposals for improvement. As Francis Rapp notes, however, the echoes of the sermon he pronounced before the synod died away quickly enough and the religious elite chose to carry on as before. There is no record of Geiler repeating this formal attempt at diocesan reform, and Rapp attributes this to the preacher's dawning awareness that Albert, the ecclesiastical linchpin of the project, was deeply cynical about the possibility of reform and enthusiastically distracted by the finances of the diocese which he inherited upon occupying the episcopate in 1478.[53]

Having failed to move his peers, Geiler turned his attention to Strasbourg's secular authorities with varying degrees of success. His agenda was wide-ranging and comprehensive, including the restoration and protection of ecclesiastical privileges, the restoration and preservation of sacred times and spaces, just legal practice in municipal courts, the restoration of true welfare, the establishment of a functioning poor relief system, and the reformation of morals. The society of this heaven-on-earth would have been tightly, if not rigidly, ordered: each member ought to remain in his or her rank, given by God at birth, and there should be no movement up or down the social hierarchy. There ought to be strict separation of religious from lay and of the sexes (women were to be either at home or in church). In the face of the apocalypse predicted for 1500 – a prophecy which Geiler made in 1499 – a lifestyle of regular repentance, penitence, and preparation for a blessed death was the model of civic life he proposed for his city. His failure to impose this model is not surprising; what draws attention is the openness of Strasbourg's *Rat* to Geiler's suggestions in specific instances, including the regulation of prostitution or the founding of an infirmary for those afflicted by the newly arrived bacteria leading to syphilis. His communication with civil authorities was conducted occasionally in writing, but for the most part, his message was delivered orally to an audience standing before the stone pulpit in the Cathedral.[54]

Concluding this brief profile of the potential members of Geiler's audience in the Cathedral and his understanding of the role of the clergy in his community, it is worth noting that he also preached in places other than the Cathedral: to women in nearby convents (including those of St. Magdalena, St. Margarethe and St. Agnes, and St. Katharina, among others), to audiences in Augsburg, and, upon invitation, to royalty. He was well-known in the German-speaking lands during his lifetime; we know this not from his boasts of his own reputation, but from the formal respect shown to him during his life: materially, as shown by the pulpit built for him, and intangibly, as demonstrated by the favour shown him by the cathedral chapter and the Emperor, as well as the respect shown by the City Councils. These combine with the accolades given him upon his death by Wimpfeling and Brant, among others, to support the historical claim of his having been held in universal high esteem. At the heart of his appeal was his talent in delivering a good sermon.[55]

The audience of a late fifteenth-century sermon could number to the thousands, with special outdoor venues being constructed for a popular preacher. In this context, the construction of a stone pulpit for Geiler in the Münster's nave does not stand out as an exaggerated gesture but, rather, one communicating unmistakeable appreciation. Strasbourg's desire to regularly enjoy his talents is more pointedly found in the creation of a permanent position and a contract, because, since the establishment of the mendicant orders, preaching had largely slipped out of the Mass liturgy to fall within their purview while also playing an important role in the wider information culture of pre-modern society. A famous preacher could anticipate receiving a formal and warm welcome upon arrival in a city; he would have negotiated a contract articulating the number of sermons, the schedule of their delivery, the location, and the remuneration he could expect from either a religious or secular employer. Following the satisfactory conclusion of his contract, he would then move on to his next engagement. Geiler's contract, however, specified that, except for his annual four weeks of leave, he was to sleep within the city walls unless given explicit permission to do otherwise by the Dean of the chapter.[56]

Audience members generally stood during the sermon, although some brought stools and benches were sometimes available for distinguished citizens or pregnant women. Some must have leaned against the wall or a pillar. Whether inside or out-of-doors, men were divided from women; when inside a church or cathedral, women were on the south side of the nave and men on the north. At sermons delivered out-of-doors, the sexes were separated by a rope. In the Strasbourg Cathedral, this division would have placed the distaff side further away from Geiler, while the men gathered around the new pulpit. A reason for this segregation might be that, according to one source, women greatly outnumbered men in the audience and the noise of their children made it difficult to hear the preacher.[57]

The cries of small children would not have been the only competition for the preacher, as audience participation in a sermon could be expected and might include heated rejoinders, jokes, arguments, or active disputation among those who were giving their attention to the pulpit – particularly if his comments were sharp. Others gave their attention elsewhere: to their pet falcons, to dogs or other animals, to laughing, strolling, and chatting with their neighbours, to attractive members of the opposite sex, or to sleep. Simply gaining the attention of the people in the nave may have been a victory for Geiler.[58]

One element of Geiler's appeal to the Strasbourgeois, which fostered his reputation and regularly attracted audience members, was his spontaneous willingness to criticize the power-holders of Strasbourg alongside the average Christian. Late in 1500, Geiler railed against the city councillors as being all on the side of the devil, like their ancestors and progeny. Like other preachers who confronted political authority, Geiler was invited to explain himself to the *Rat*. Unlike other preachers, however, his stature in Strasbourg and his long-term commitment to the well-being of its indwellers had earned him time to justify and explain: two councillors met him in the cathedral (his chosen meeting place) to discuss the matter. On 27 January 1501, he appeared before the Senate and XXI to voice his complaints in person and on 27 March, submitted a copy in writing. The XXI Articles of Complaint were moderate and aligned with existing legal structure; in the following years, some of his proposals were accepted and put into practice. The respect accorded Geiler from Strasbourg's city council was rare; while preachers may have regularly attempted to awaken the Christian consciences of civil authorities, imprisonment, banishment, or exile was a more frequent consequence.[59]

A sermon in the late medieval *Liebfrauenmünster zu Straßburg* combined moral instruction with entertainment, allowed for socializing in a lightly regulated venue, and invited the humblest to experience a taste of heavenly egalitarianism through communal worship alongside the most exalted in the land. A preacher's ambition was to capture and hold the attention of all who came, regardless of their motivation for being there. Geiler not only succeeded in doing so for three decades, but inspired such affection among the cathedral-goers that he continued to be present among them after his death, as it was felt appropriate to bury him at the foot of the stone pulpit he occupied for 32 years. His representation of the natural world was unexceptional to his community; they took his views for granted as reasonable and likely true.

This chapter has provided a brief description of the physical environment of Alsace and a summary of current knowledge of the region's climate in the sixteenth century. As well, I introduced the sources from which I built the weather report featured in Chapters 2–4, and sketched an outline of social conditions and relations in Strasbourg and Alsace during the late fifteenth and early sixteenth centuries. This last supplies a rough idea of the social setting into which the orthodox Roman Catholic, radical, and Reformed representations of the natural world were delivered, as well as a basis of comparison for changes in those settings and views after 1517. The different views of Geiler, Zyegler, and Capito are topics for the next three chapters.

Notes

1 Much excellent research has been conducted on Reformation-era Strasbourg and Alsace, and my debt for this section is owed to the following works of scholarship, presented in alphabetical order for convenience: Martin Allheilig, ed., *Artisans et ouvriers d'Alsace*, Publications de la Société Savante d'Alsace et des Régions de l'Est, IX (Strasbourg: Librairie Istra, 1965); George Bischoff, *La Guerre des Paysans*; Peter Blickle, *From the Communal Reformation to the Revolution of the Common Man*, trans. Beat Kümin (Leiden: Brill, 1998); Peter Blickle, *The Revolution of 1525: The German Peasants' War from a New Perspective*, trans. Thomas A. Brady, Jr., and H.C. Erik Midelfort (Baltimore and London: The Johns Hopkins University Press, 1985); Thomas A. Brady, Jr., *Ruling Class, Regime and Reformation at Strasbourg, 1520–1555* (Leiden: Brill, 1978); Thomas A. Brady, Jr., *German Histories in the Age of Reformations: 1400–1650* (Cambridge: Cambridge University Press, 2009); Thomas A. Brady, Jr., Heiko A. Oberman, and James D. Tracy, *Handbook of European History, 1400–1600: Late Middle Ages, Renaissance and Reformation*, 2 vols. (Leidon: Brill, 1994); Miriam Usher Chrisman, *Strasbourg and the Reform: A Study in the Process of Change* (New Haven and London: Yale University Press, 1967); Marguerite Parayre-Kuntzel, *L'Église et la vie quotidienne du paysan d'Alsace au Moyen Age* (Strasbourg: Librairie Istra, 1975); Francis Rapp, 'The Social and Economic Prehistory of the Peasant War in Lower Alsace,' in *The German Peasant War of 1525 – New Viewpoints*, eds. and trans. Bob Scribner and Gerhard Benecke (London: George Allen & Unwin, 1979); Tom Scott, *Freiburg and the Breisgau: Town-Country Relations in the Age of Reformation and Peasants' War* (Oxford: Clarendon Press, 1986); Tom Scott, *Regional Identity and Economic Change: The Upper Rhine 1450–1600* (Oxford:

Clarendon Press, 1997); Tom Scott, *The Early Reformation in Germany: Between Secular Impact and Radical Vision* (Farnham: Ashgate, 2013); and Charlotte A. Stanford, *Commemorating the Dead in Late Medieval Strasbourg: The Cathedral's Book of Donors and Its Use (1320–1521)* (Farnham: Ashgate, 2011).

2 Bischoff, 33–7; Tom Scott, *Regional Identity and Economic Change*, 17–40.

3 Roger G. Barry, *Mountain Weather and Climate*, 2nd Edition (London: Routledge, 2013), Ch. 4; See, for example, Arthur R. Kruckeberg, *Geology and Plant Life: The Effects of Landforms and Rock Types on Plants* (Washington: University of Washington Press, 2004).

4 Christian Röhr, 'Lage des Oberrheingrabens in einer Karte von Deutschland',' *Der Oberrheingrabe*, modified from Lahner, L. and M. Toloczyki, *Geowissenschaftliche Karte der Bundesrepublik Deutschland*, 2006 <http://www.oberrhein-graben.de/index.htm> [page last updated 23 October 2014; accessed 18 October 2023]; Gideon G.O. Lopez Cardozo and Jan H. Behrmann, 'Kinematic Analysis of the Upper Rhine Graben Boundary Fault System',' *Journal of Structural Geology*, 28 (2006), 1028–39; Wantzen, Karl M., Urs Ühlinger, Gerald Van der Velde, Rob S.E.W. Leuven, Laurent Schmitt, and Jean-Nicolas Beisel, 'The Rhine River Basin',' in *Rivers of Europe*, eds. Klement Tockner, Christiane Zarfl, and Christopher T. Robinson, 2nd Edition (Amsterdam: Elsevier Ltd., 2022), 333–91.

5 Scott, *Regional Identity and Economic Change*, 20; B. Forel et al., 'Historical Mining and Smelting in the Vosges Mountains (France) Recorded in Two Ombrotrophic Peat Bogs,' *Journal of Geochemical Exploration*, 107 (2010), 9–20; Rudiger Mäckel, Arne Friedmann, and Dirk Sudhaus, 'Environmental Changes and Human Impact on Landscape Development in the Upper Rhine Region',' *Erdkunde*, 6 (2009), 43 and 45; see also Forel et al., 11; Pierre Fluck, J. Gauthier, and A. Disser, 'The Alsatian Altenberg: A Seven Centuries Laboratory for Silver Metallurgy',' *Archéomètallurgie en Europe*, Troisième Conférence Internationale, France, 2011; available online at HAL (*Hyper Articles en Ligne*), https://hal.science/hal-00914305/document doi: hal- 00914305, version 1 [accessed 10 January 2024]; P. Hestin, deep-pit guide at Tellure, Ste-Marie-aux-Mines (Personal communication, 28 March 2013).

6 Scott, *Regional Identity and Economic Change*, 25.

7 Kamil Ustaszewski and Stefan M. Schmid, 'Control of Preexisting Faults on Geometry and Kinematics in the Northernmost Part of the Jura Fold-and-Thrust Belt,' *Tectonics*, 25 TC5003, doi: 10.1029/2005TC001915.

8 Magdala Tesauro et al., 'Continuous GPS and Broad-Scale Deformation across the Rhine Graben and the Alps,' *International Journal of Earth Sciences*, 94 (2005), 525–6.

9 Fritz Oehl et al., 'Soil Type and Land Use Intensity Determine the Composition of Arbuscular Mycorrhizal Fungal Communities,' *Soil Biology and Biochemistry*, 42 (2010), 724–38.

10 A. Lang et al., 'Changes in Sediment Flux and Storage within a Fluvial System: Some Examples from the Rhine Catchment,' *Hydrological Processes*, 17(2003), 3321–34 (3322); Eva Johanna Link, 'Investigation and Modeling of the Optimization Potential of Adapted Nitrogen Fertilization Strategies in Corn Cropping Systems with Regard to Minimize Nitrogen Losses,' thesis for a Doctor of Agricultural Sciences, Universität Hohenheim, Hohenheim, 2005, 17 <http://opus.uni-hohenheim.de/volltexte/2005/122/pdf/2005-11-30_Dissertation_JLink.pdf> [accessed 30 October 2023]; From textual analysis, the Upper Rhine Revolutionary is thought to have been a man; his work is very likely to have reached completion in 1509 or 1510. Der Oberrheinische Revolutionär, *das Buchli der hundert Capiteln mit XXXX Statuten*, ed. Klaus H. Lauterbach (Hannover: Hahn, 2009).

11 Marc Cioc, 'The Political Ecology of the Rhine,' in *Nature in German History*, ed. Christof Mauch, Studies in German History 1 (New York and Oxford: Berghahn, 2004), 31–47.

12 Koos Wieriks and Anne Schulte-Wülwer-Leidig, 'Integrated Water Management for the Rhine River Basin, from Pollution Prevention to Ecosystem Improvement,' *Natural Resources Forum*, 21 (1997), 147–56; Lang, 3322.

13 Angelika Sadlau et al., *Die lange Bruck: 600 Jahre Wege zum Nachbarn* (Kehl: Kehl am Rhein Kultur- un Verkehrsamt, 1989), 39.

14 Jean-Paul Haettel and Edmond Maennel, *Strasbourg et ses Ponts* (Strasbourg: Le Verger Éditeur, 1990).

15 R. Ostermann and A. Reif, 'Socioeconomical and Ecological Aspects of Coppice Woods History in the Lower Vosges (France) and the Black Forest (Germany)',' in *Methods and Approaches in Forest History*, eds. Mauro Agnoletti and Steven Anderson, International Union of Forestry Research Organizations (IUFRO) research series 3 (Wallingford: CABI, 2000), 107–18.

16 Mäckel et al., 45.

17 Christian Pfister et al., 'Documentary Evidence on Climate in Sixteenth-Century Europe,' *Climatic Change*, 43 (September 1999), 58.

18 Mauelshagen, 39–42.

19 Sebastian Franck, *Chronica, Zeytbuch vnd geschychtbibel von anbegyn biß inn diß gegenwertig MDXXXI. jar: Darin beide Gottes vnd der welt lauff, hendel, art, wort, werck, thun, lassen, kriegen, wesen vnd leben ersehen vnd begriffen wirt ...* (Straßburg: Balthassar Beck, 1531).

20 Oliver Wetter and Christian Pfister, 'Spring-Summer Temperatures Reconstructed for Northern Switzerland and Southwestern Germany from Winter Rye Harvest Dates, 1454–1970',' *Climate of the Past,* 7 (2011), 1307–26.

21 Moving averages are commonly used to model data sets with only one variable; they serve to mitigate the effect of errors or random shocks on the averages. A±2 × Sigma threshold means that 95% of the values lay within two standard units of the average.

22 Stefan Brönnimann, 'From Climate to Weather Reconstructions,' *PLOS Climate*, 1(6), e0000034, 1, https://doi.org/10.1371/journal. Pclm.0000034.

23 Christian Pfister et al., 'Documentary Evidence as Climate Proxies. White Paper written for the Proxy Uncertainty Workshop in Trieste, 9–11 June 2008 (updated December 2008)',' National Centers for Environmental Information (USA), https://www.ncei.noaa.gov/pub/data/paleo/historical/documentary.pdf [accessed 19 October 2023]; After Strasbourg's library had been bombed, Abbott Léon Dacheux and Rodolphe Reuss gathered the remaining fragments and published them in a four-volume series from 1887 to 1901. They are: Sébald Büheler, *I. La Petite Chronique de la Cathêdrale. La Chronique Strasbourgeoise de Sébald Büheler*, ed. Léon Dacheux, Fragments des Anciennes Chroniques d'Alsace I (Strasbourg: R. Schultz & Cⁱᵉ, 1887); Daniel Specklin, *II. Les Collectanées de Daniel Specklin, Chronique Strasbourgeoise du seizième siècle*, ed. Rodolphe Reuss, Fragments des Anciennes Chroniques d'Alsace II (Strasbourg: Librairie J. Noiriel, 1890); Jacques Trausch, Jean Wencker, Sébastien Brant, *III. Les Chroniques Strasbourgeoises de Jacques Trausch et de Jean Wencker. IV. Les Annales de Sébastien Brant*, ed. Léon Dacheux, Fragments des Anciennes Chroniques d'Alsace III (Strasbourg: R. Schultz & Cⁱᵉ, 1892); Jacob Twinger von Königshoven, Materne Berler, anonymous authors, and Sébastien Brant, *VII. Koenigshoven. Fragments de la Chronique Latine. VIII. Berler. Fragments de la Chronique. IX. Fragments de Diverses Vieilles Chroniques. X. Les Annales de Sébastien Brant*, ed. Léon Dacheux, Fragments des Anciennes Chroniques d'Alsace IV (Strasbourg: R. Schultz & Cⁱᵉ, 1901).

Jacob Twinger von Königshoven's chronicle from the early fifteenth century was included in another collection edited by Karl Hegel.

24 Johann Jacob Meyer, *La Chronique Strasbourgeoise*, ed. Rodolphe Reuss (Strasbourg: J. Noiriel, 1873); Frère Martin Stauffenberger, 'Les Annales des Frères Mineurs de Strasbourg, rédigées par le frère Martin Stauffenberger, économe du couvent (1507–1510)',' in *Bulletin de la Société pour la Conservation des Monuments Historiques*, ed. Rodolphe Reuss, IIe Serie, 18 (Strasbourg: R. Schultz & Cie, 1897), 295–314; Jacques de Gottesheim, 'Les Éphémérides de Jacques de Gottesheim: Docteur en Droit, Prébendier du Grand-Choeur de la Cathédrale (1524–1543). Fragments publiés pour la première fois et annotés',' *Bulletin de la Société pour la Conservation des Monuments Historiques*, ed. Rodolphe Reuss, IIe Serie, XIX, bk. I (Strasbourg: R. Schultz & Cie, 1898), 261–81; Daniel Specklin, *Les Collectanées de Daniel Specklin, Chronique Strasbourgeoise du seizième siècle.*, ed. Rodolphe Reuss, Fragments des Anciennes Chroniques d'Alsace II (Strasbourg: J. Noiriel, 1890); Franz-Josef Mone, ed., 'Fortsetzungen des Königshoven. Straßburger Zusätze',' in *Quellensammlung der badischen Landesgeschichte*, 1 (Karlsruhe: C. Macklot, 1848), 252–80; Franz-Josef Mone, ed., 'Straßburger Jahrgeschichten (1424–1593)',' in *Quellensammlung der badischen Landesgeschichte*, 2 (Karlsruhe: C. Macklot, 1848), 138–45; P.F. Malachias Tschamser, *Annales oder Jahrs-Geschichten der Baarfüseren oder Minderen Brüdern S. Franc. Ord. inßgemein Conventualen genannt, zu Thann*, I and II (Colmar: Hoffmann, 1864). For more information about Tschamser and his *Annales*, see Paul Weiss, *1516–1700, heurs et malheurs d'une ville et d'une province: selon les „Chroniques" de Malachie Tschamser, père-gardien des Franciscains de Thann* (Boofzheim: ACM Édition, 2000); Xavier Mossmann, ed., *Chronique des dominicains de Guebwiller* (Guebwiller: G. Brückert; Colmar: L. Reiffinger; Strasbourg: Schmidt et Grucker, 1844).

25 Meyer, 9–10.

26 Stauffenberger, 296.

27 Gottesheim, 261–5. The copyist declares that he has seen the legal document which established Gottesheim in the role, which was held by the episcopal notary Michel Schwencker von Gerusbach. Ruess speculates that Gottesheim may have been born c. 1490 in order to hold this post.

28 For more information on Specklin's life, see Albert Fischer, *Daniel Specklin aus Strassburg (1536–1589): Festungsbaumeister, Ingenieur und Kartograph* (Sigmaringen: Jan Thorbecke Verlag, 1996); Specklin, 9–10.

29 This paragraph is based on Hansmartin Schwarzmaier, 'Mone, Franz Joseph,' *Neue Deutsche Biographie*, 18 (1997), 32 f, <http://www.deutsche-biographie.de/pnd118861476.html>, accessed 19 October 2023.

30 Mossmann, Dedication, 3; Hans Stolz, *Die Hans Stolz'sche Gebweiler Chronik. Zeugenbericht über den Bauernkrieg am Oberrhein*, ed. Wolfram Stolz (Freiburg: Edition Stolz, 1979).

31 Jean-Pierre Kintz, 'Notes sur Quelques Aspects Demographiques de la Ville de Strasbourg',' in *Strasbourg au Coeur Religieux du XVIe Siècle*, eds. Georges Livet and Francis Rapp, Société Savante d'Alsace et des Régions de L'est, Collection 'Grandes publication',' XII (Strasbourg: Librairie Istra, 1977), 13; Martin Greschat, *Martin Bucer: A Reformer and His Times*, trans. Stephen E. Buckwalter (Louisville and London: Westminster John Knox Press, 2004), 2; Hajo Holborn, *A History of Modern Germany: The Reformation*, Vol. 1 (Princeton and Oxford: Princeton University Press, 1982), 37–8; Andrew Pettegree, *Europe in the Sixteenth Century* (Oxford: Blackwell Publishing, 2002), 1. See also Bischoff, 50. Estimates of the population of Alsace at that time are not yet available. While the subdiscipline of environmental history frequently includes the bacterial and viral environments,

those are not the focus of this work and will be touched on only lightly. A good synopsis of the debate on urban mortality is available in Robert Woods, 'Urban-Rural Mortality Differentials: An Unresolved Debate',' *Population and Development Review*, 29 (2003), 29–46.

32 Thomas A. Brady, Jr., *German Histories*, 143; Bischoff, 58.
33 Johannes Geiler von Kaysersberg, *Die Emeis*, ed. Johannes Pauli (Strasbourg: Johann Grüninger, 1517), fol VIII^v.
34 For an exhaustive description of political and economic conditions during the Bundschuh period in Alsace (1493–1525), see Georges Bischoff's *La guerre des Paysans*.
35 Bischoff, 45–8.
36 Philippe Dollinger, 'L'évolution politique des corporations strasbourgeoises à la fin du moyen âge',' in *Artisans et ouvriers d'Alsace*, ed. Hans Haug, Publications de la Société Savante d'Alsace et des Régions de l'Est, IX (Strasbourg: Librairie Istra, 1965), 127–133; Brady, *Protestant Politics*, 17. See also R. Po-Chia Hsia, 'The Myth of the Commune: Recent Historiography on City and Reformation in Germany,' *Central European History*, 20 (1987), 209.
37 Brady, *Ruling Class, Regime and Reformation*, 118, 163–5 and 171. The reason for the numeric titles is unclear, since the number of councillors grew and diminished over time.
38 Debra Kaplan, *Beyond Expulsion: Jews, Christians, and Reformation Strasbourg*, Stanford Studies in Jewish History and Culture (Palo Alto: Stanford University Press, 2011), 69–70.
39 Brady, *Ruling Class, Regime and Reformation*, 38–41.
40 Brady, *Ruling Class, Regime and Reformation*, 174–8; Bischoff, 71 and 253; Scott, *Freiburg and the Breisgau*, 18; Brady, *Protestant Politics*, 16; Bischoff, 51; Scott, *Regional Identity and Economic Change*, 201–72.
41 Brady, *Ruling Class, Regime and Reformation*, 96.
42 Greschat, 6.
43 Stanford, 197–205.
44 Stanford, 26 and 35–8; Philip 1 of France (1076–1093) established the *Pfund-Pfenning*, also known in Strasbourg as *feine Marck* or *Livres deniers*, to be measures of eight ounces of pure silver alloy – although intervening centuries saw purity of greater and lesser degree. Louis Levrault, *Essai sur l'ancienne monnaie de Strasbourg et sur ses rapports avec l'histoire de la Ville et de l'Évêché* (Strasbourg and Paris: Levrault et Bertrand, 1842), 205–7.
45 Rapp, 57; Markus Cerman, *Villages and Lords in Eastern Europe, 1300–1800*, Studies in European History (Palgrave Macmillan, 2012).
46 Blickle, *The Revolution of 1525*, 50; Bischoff, 277.
47 Bischoff, 64–5; Rapp, 53; Bischoff, 69 and 277.
48 Rapp, 52–62; Bischoff, 282–9 and 268; Scott, *Regional Identity and Economic Change*, 110.
49 Bischoff, 451–2; Blickle, *The Revolution of 1525*, 268; Scott, *Regional Identity and Economic Change*, 201 and 310; Bischoff, 61–3. *Landsknechte* were mercenary foot soldiers, often forming the bulk of the army of the Holy Roman Empire during this period.
50 Scott, *Regional Identity and Economic Change*, 203, 287–322; Gadi Algazi, 'Lords Ask, Peasants Answer: Making Traditions in Late Medieval Village Assemblies,' in *Between History and Histories: The Making of Silences and Commemorations* (Toronto: University of Toronto Press, Anthropological Horizons Series, 1997), 199–229.
51 Rita Voltmer, *Wie der Wächter auf dem Turm: ein Prediger und seine Stadt Johannes Geiler von Kaysersberg (1445–1510) und Straßburg*, Beiträge zur Landes- und

Kulturgeschichte, 4, ed. Franz Irsigler (Trier: Porta Alba, 2005), 547–82; quoted in Bischoff, 62; Jean-Pierre Kintz, 'Strasbourg cité refuge: mendiants, fugitifs, exiles',' in *Regards sur l'histoire de l'Alsace XVIe- XXe siècle*, Hommage de la Fédération des Sociétés d'Histoire et d'Archéologie au Professeur Jean- Pierre Kintz (Drulingen: Scheuer, 2008), 21.

52 Geiler, fol VIII^v and fol. IX^v.

53 Rita Voltmer, 'Political Preaching and a Design of Urban Reform: Johannes Geiler of Kayserberg and Strasbourg',' *Franciscan Studies*, 71 (2013), 71–87; Francis Rapp, 'Jean Geiler de Kaysersberg (1445–1510): le prédicateur de la Cathédrale de Strasbourg',' in *Grandes figures de l'humanisme Alsacien: Courants Milieux Destins*, eds. Francis Rapp and Georges Livet, Publications de la Société savante d'Alsace et des régions de l'Est, Collection 'Grandes Publications',' XIII (Strasbourg: Istra, 1978), 25–32 (30).

54 Voltmer, 'Political Preaching,' 83–4 and 86; Voltmer, *Wie der Wächter*, 122–4.

55 Voltmer, *Wie der Wächter*, 944, 950, 951, and 961; Geiler was a chaplain for Maximilian I, and, as mentioned above, both occasionally preached to the Emperor and at least once enjoyed his confidence. E. Jane Dempsey Douglass, *Justification in Late Medieval Preaching: A Study of John Geiler of Keisersberg* (Leiden: Brill, 1966), 5; Brady, *German Histories in the Age of Reformation*, 140.

56 Larissa Taylor, *Soldiers of Christ: Preaching in Late Medieval and Reformation France* (Toronto: University of Toronto Press, 2002), 9, 21, and 28–9. While Taylor's primary sources are from France, she recognizes that much of her theoretical inspiration and secondary source material is in reference to Germany. It does not seem inappropriate to take advantage of her work on preaching for my study of Alsace, particularly since she refers to Geiler at several instances. Andrew Pettegree, *The Reformation and the Culture of Persuasion* (Cambridge: Cambridge University Press, 2005), 10 and 21.

57 Hervé Martin, *La metier du prédicateur* à *la fin du Moyen Age* (Paris: Cerf, 1988), 558.

58 Taylor, 33–4.

59 Johannes Geiler von Kaisersberg, *Die æltesten Schriften Geiler von Kaysersberg*, ed. Léon Dacheux (Freiburg im Breisgau: Herderische Verlagshandlung, 1877), 1; Taylor, 35; Voltmer, *Wie der Wächter*, 418–32; Charles Guillaume A. Schmidt, *Histoire Littéraire de L'Alsace a la fin du XVe et au commencement du XVIe siècle*, 2 vols (Hildesheim: Georg Ohns Verlagsbuchhandlung, 1966; orig. France: Sandoz et Fischbacher, 1879), 362–5.

References

Primary sources

Büheler, Sébald, *I. La Petite Chronique de la Cathêdrale. La Chronique Strasbourgeoise de Sébald Büheler*, ed. Léon Dacheux, Fragments des Anciennes Chroniques d'Alsace I (Strasbourg: R. Schultz & Cie, 1887).

Der Oberrheinische Revolutionär, *das Buchli der hundert Capiteln mit XXXX Statuten*, ed. Klaus H. Lauterbach (Hannover: Hahn, 2009).

Franck, Sebastian, *Chronica, Zeytbuch vnd geschychtbibel von anbegyn biß inn diß gegenwertig MDXXXI. jar: Darin beide Gottes vnd der welt lauff, hendel, art, wort, werck, thun, lassen, kriegen, wesen vnd leben ersehen vnd begriffen wirt…* (Straßburg: Balthassar Beck, 1531).

Geiler von Kaysersberg, Johannes, *Die Emeis*, ed. Johannes Pauli (Strasbourg: Johann Grüninger, 1517).

———, *Die ältesten Schriften Geiler von Kaysersberg*, ed. Léon Dacheux (Freiburg im Breisgau: Herderische Verlagshandlung, 1877).

Gottesheim, Jacques de, 'Les Éphémérides de Jacques de Gottesheim: Docteur en Droit, Prébendier du Grand-Choeur de la Cathédrale (1524–1543). Fragments publiés pour la première fois et annotés,' in *Bulletin de la Société pour la Conservation des Monuments Historiques*, ed. Rodolphe Reuss, IIᵉ Serie, XIX, bk. I (Strasbourg: R. Schultz & Cie, 1898), 261–81.

Meyer, Johann Jacob, *La Chronique Strasbourgeoise*, ed. Rodolphe Reuss (Strasbourg: J. Noiriel, 1873).

Mone, Franz-Josef, ed., 'Fortsetzungen des Königshoven. Straßburger Zusätze,' in *Quellensammlung der badischen Landesgeschichte*, 1 (Karlsruhe: C. Macklot, 1848), 252–80.

———, 'Straßburger Jahrgeschichten (1424–1593),' in *Quellensammlung der badischen Landesgeschichte*, 2 (Karlsruhe: C. Macklot, 1848), 138–45.

Mossmann, Xavier, ed., *Chronique des dominicains de Guebwiller* (Guebwiller: G. Brückert; Colmar: L. Reiffinger; Strasbourg: Schmidt et Grucker, 1844).

Specklin, Daniel, *II. Les Collectanées de Daniel Specklin, Chronique Strasbourgeoise du seizième siècle*, ed. Rodolphe Reuss, Fragments des Anciennes Chroniques d'Alsace II (Strasbourg: Librairie J. Noiriel, 1890).

Stauffenberger, Frère Martin, 'Les Annales des Frères Mineurs de Strasbourg, rédigées par le frère Martin Stauffenberger, économe du couvent (1507–1510),' in *Bulletin de la Société pour la Conservation des Monuments Historiques*, ed. Rodolphe Reuss, IIᵉ Serie, 18 (Strasbourg: R. Schultz & Cie, 1897), 295–314.

Stolz, Hans, *Die Hans Stolz'sche Gebweiler Chronik. Zeugenbericht über den Bauernkrieg am Oberrhein*, ed. Wolfram Stolz (Freiburg: Edition Stolz, 1979).

Trausch, Jacques, Jean Wencker, Sébastien Brant, *III. Les Chroniques Strasbourgeoises de Jacques Trausch et de Jean Wencker. IV. Les Annales de Sébastien Brant*, ed. Léon Dacheux, Fragments des Anciennes Chroniques d'Alsace III (Strasbourg: R. Schultz & Cᵢᵉ, 1892).

Tschamser, P.F. Malachias, *Annales oder Jahrs-Geschichten der Baarfüseren oder Minderen Brüdern S. Franc. Ord. inßgemein Conventualen genannt, zu Thann*, I and II (Colmar: Hoffmann, 1864).

Twinger von Königshoven, Jacob, Materne Berler, anonymous authors, and Sébastien Brant, *VII. Koenigshoven. Fragments de la Chronique Latine. VIII. Berler. Fragments de la Chronique. IX. Fragments de Diverses Vieilles Chroniques. X. Les Annales de Sébastien Brant*, ed. Léon Dacheux, Fragments des Anciennes Chroniques d'Alsace IV (Strasbourg: R. Schultz & Cᵢᵉ, 1901).

Secondary sources

Algazi, Gadi, 'Lords Ask, Peasants Answer: Making Traditions in Late Medieval Village Assemblies,' in *Between History and Histories: The Making of Silences and Commemorations*, eds. Gerald Sider and Gavin Smith, Anthropological Horizons 11 (Toronto: University of Toronto Press, Anthropological Horizons Series, 1997), 199–229.

Allheilig, Martin, ed., *Artisans et ouvriers d'Alsace*, Publications de la Société Savante d'Alsace et des Régions de l'Est, IX (Strasbourg: Librairie Istra, 1965).

Barry, Roger G., *Mountain Weather and Climate*, 2nd Edition (London: Routledge, 2013).

Bischoff, Georges, *La Guerre des Paysans. L'Alsace et la révolution du Bundschuh 1493–1525* (Strasbourg: Éditions La Nuée Bleue, 2010).

Blickle, Peter, *From the Communal Reformation to the Revolution of the Common Man*, trans. Beat Kümin (Leiden: Brill, 1998).

———, *The Revolution of 1525: The German Peasants' War from a New Perspective*, trans. Thomas A. Brady, Jr., and H.C. Erik Midelfort (Baltimore and London: The Johns Hopkins University Press, 1985).

Brady, Thomas A., Jr., *Ruling Class, Regime and Reformation at Strasbourg, 1520–1555* (Leiden: Brill, 1978).

———, *German Histories in the Age of Reformations: 1400–1650* (Cambridge: Cambridge University Press, 2009).

Brady, Thomas A., Jr., Heiko A. Oberman, and James D. Tracy, *Handbook of European History, 1400–1600: Late Middle Ages, Renaissance and Reformation*, 2 vols. (Leidon: Brill, 1994).

Brönnimann, Stefan, 'From Climate to Weather Reconstructions,' *PLOS Climate*, 1(6) (2022), e0000034, doi: 10.1371/journal. Pclm.0000034.

Cardozo, Gideon G.O. Lopez, and Jan H. Behrmann, 'Kinematic Analysis of the Upper Rhine Graben Boundary Fault System,' *Journal of Structural Geology*, 28 (2006), 1028–39.

Chrisman, Miriam Usher, *Strasbourg and the Reform: A Study in the Process of Change* (New Haven, CT and London: Yale University Press, 1967).

Cerman, Markus, *Villages and Lords in Eastern Europe, 1300–1800*, Studies in European History (Basingstoke: Palgrave Macmillan, 2012).

Cioc, Marc, 'The Political Ecology of the Rhine,' in *Nature in German History*, ed. Christof Mauch, Studies in German History 1 (New York and Oxford: Berghahn, 2004), 31–47.

Dollinger, Phillippe, 'L'évolution politique des corporations strasbourgeoises à la fin du moyen âge,' in *Artisans et ouvriers d'Alsace*, ed. Hans Haug, Publications de la Société Savante d'Alsace et des Régions de l'Est, IX (Strasbourg: Librairie Istra, 1965), 127–133.

Douglass, E. Jane Dempsey, *Justification in Late Medieval Preaching: A Study of John Geiler of Keisersberg* (Leiden: Brill, 1966).

Fischer, Albert, *Daniel Specklin aus Strassburg (1536–1589): Festungsbaumeister, Ingenieur und Kartograph* (Sigmaringen: Jan Thorbecke Verlag, 1996).

Forel, B., F. Monna, C. Petit, O. Bruguier, R. Losno, P. Fluck, C. Begeot, H. Richard, V. Bichet, and C. Chateau, 'Historical Mining and Smelting in the Vosges Mountains (France) Recorded in Two Ombrotrophic Peat Bogs,' *Journal of Geochemical Exploration*, 107 (2010), 9–20.

Fluck, Pierre, J. Gauthier, and A. Disser, 'The Alsatian Altenberg: A Seven Centuries Laboratory for Silver Metallurgy,' *Archéomètallurgie en Europe*, Troisième Conférence Internationale, France 2011, available online at HAL (Hyper Articles en Ligne), https://hal.science/hal-00914305/document, doi: hal-00914305, version 1 [accessed 10 January 2024].

Greschat, Martin, *Martin Bucer: A Reformer and His Times*, trans. Stephen E. Buckwalter (Louisville and London: Westminster John Knox Press, 2004).

Haettel, Jean-Paul, and Edmond Maennel, *Strasbourg et ses Ponts* (Strasbourg: Le Verger Éditeur, 1990).

Holborn, Hajo, *A History of Modern Germany: The Reformation*, Vol. 1 (Princeton and Oxford: Princeton University Press, 1982), 37–8.

Hsia, R. Po-Chia, 'The Myth of the Commune: Recent Historiography on City and Reformation in Germany,' *Central European History*, 20 (1987), 203–15.

Kaplan, Debra, *Beyond Expulsion: Jews, Christians, and Reformation Strasbourg*, Stanford Studies in Jewish History and Culture (Palo Alto: Stanford University Press, 2011).

Kintz, Jean-Pierre, 'Notes sur Quelques Aspects Demographiques de la Ville de Strasbourg,' in *Strasbourg au Coeur Religieux du XVIe Siècle*, eds. Georges Livet and Francis Rapp, Société Savante d'Alsace et des Régions de L'est, Collection 'Grandes publication,' XII (Strasbourg: Librairie Istra, 1977).

———— 'Strasbourg cité refuge: mendiants, fugitifs, exiles,' in *Regards sur l'histoire de l'Alsace XVIe-XXe siècle*, Hommage de la Fédération des Sociétés d'Histoire et d'Archéologie au Professeur Jean- Pierre Kintz (Drulingen: Scheuer, 2008), 20–48.

Kruckeberg, Arthur R., *Geology and Plant Life: The Effects of Landforms and Rock Types on Plants* (Washington: University of Washington Press, 2004).

Lang, Andreas, Hans-Rudolf Bork, Rüdiger Mäckel, Nicholas Preston, Jürgen Wunderlich, Richard Dikau, 'Changes in Sediment Flux and Storage within a Fluvial System: Some Examples from the Rhine Catchment,' *Hydrological Processes*, 17:16 (2003), 3321–34 (3322).

Levrault, Louis, *Essai sur l'ancienne monnaie de Strasbourg et sur ses rapports avec l'histoire de la Ville et de l'Évêché* (Strasbourg and Paris: Levrault et Bertrand, 1842).

Link, Eva Johanna, 'Investigation and Modeling of the Optimization Potential of Adapted Nitrogen Fertilization Strategies in Corn Cropping Systems with Regard to Minimize Nitrogen Losses,' thesis for a Doctor of Agricultural Sciences, Universität Hohenheim, Hohenheim, 2005, 17 <http://opus.uni-hohenheim.de/volltexte/2005/122/pdf/2005-11-30_Dissertation_JLink.pdf> [accessed 30 October 2023].

Mäckel, Rudiger, Arne Friedmann, and Dirk Sudhaus, 'Environmental Changes and Human Impact on Landscape Development in the Upper Rhine Region,' *Erdkunde*, 6 (2009), 35–49.

Martin, Hervé, *La metier du prédicateur* à *la fin du Moyen Age* (Paris: Cerf, 1988).

Parayre-Kuntzel, Marguerite, *L'Église et la vie quotidienne du paysan d'Alsace au Moyen Age* (Strasbourg: Librairie Istra, 1975).

Pettegree, Andrew, *Europe in the Sixteenth Century* (Oxford: Blackwell Publishing, 2002).

Oehl, Fritz, Endre Laczko, Arno Bogenrieder, Karl Stahr, Robert Bösch, Marcel van der Heijden, and Ewald Sieverding, 'Soil Type and Land Use Intensity Determine the Composition of Arbuscular Mycorrhizal Fungal Communities,' *Soil Biology and Biochemistry*, 42:5 (2010), 724–38.

Ostermann, P., and A. Reif, 'Socioeconomical and Ecological Aspects of Coppice Woods History in the Lower Vosges (France) and the Black Forest (Germany),' *Methods and Approaches in Forest History*, eds. Mauro Agnoletti and Steven Anderson, International Union of Forestry Research Organizations (IUFRO) research series 3 (Wallinford: CABI, 2000), 107–18.

Pettegree, Andrew, *The Reformation and the Culture of Persuasion* (Cambridge: Cambridge University Press, 2005).

Pfister, Christian, Rudolf Brázdil, Rüdiger Glaser, Mariano Barriendos, Dario Camuffo, Mathias Deutsch, Petr Dobrovolný, Silvia Enzi, Emanuela Guidoboni, Oldřich Kotyza, Stefan Militzer, Lajos Rácz, and Fernando S. Rodrigo, 'Documentary Evidence on Climate in Sixteenth-Century Europe,' *Climatic Change*, 43 (September 1999), 55–110.

Pfister, Christian, Jürg Luterbacher, Heinz Wanner, Dennis Wheeler, Rudolf Brázdil, Quansheng Ge, Zhixin Hao, Anders Moberg, Stefan Grab, and Maria Rosario del Prieto, 'Documentary Evidence as Climate Proxies. White Paper Written for the Proxy Uncertainty Workshop in Trieste, 9–11 June 2008 (updated December 2008),' National Centers for Environmental Information (USA), https://www.ncei. noaa.gov/pub/data/paleo/historical/documentary.pdf [accessed 19 October 2023].

Rapp, Francis, 'Jean Geiler de Kaysersberg (1445–1510): le prédicateur de la Cathédrale de Strasbourg,' in *Grandes figures de l'humanisme Alsacien: Courants Milieux Destins*, eds. Francis Rapp and Georges Livet, Publications de la Société savante d'Alsace et des régions de l'Est, Collection 'Grandes Publications,' XIII (Strasbourg: Istra, 1978), 25–32.

——— 'The Social and Economic Prehistory of the Peasant War in Lower Alsace,' in *The German Peasant War of 1525 – New Viewpoints*, eds. and trans. Bob Scribner and Gerhard Benecke, Routledge Library Editions: Political Protest (London: George Allen & Unwin, 1979), 52–62.

Röhr, Christian, 'Lage des Oberrheingrabens in einer Karte von Deutschland,' *Der Oberrheingrabe*, modified from Lahner, L. and M. Toloczyki, *Geowissenschaftliche Karte der Bundesrepublik Deutschland*, 2006 <http://www.oberrheingraben. de/index.htm> [page last updated 23 October 2014; accessed 18 October 2023].

Sadlau, Angelika, Helmut Schneider, and Carl Helmut Steckner, *Die lange Bruck: 600 Jahre Wege zum Nachbarn* (Kehl: Kehl am Rhein Kultur- un Verkehrsamt, 1989).

Schmidt, Charles Guillaume A., *Histoire Littéraire de L'Alsace a la fin du XVe et au commencement du XVIe siècle*, 2 vols (Hildesheim: Georg Ohns Verlagsbuchhandlung, 1966; orig. France: Sandoz et Fischbacher, 1879).

Schwarzmaier, Hansmartin, 'Mone, Franz Joseph,' in *Neue Deutsche Biographie* 18 (1997), 32 f, <http://www.deutsche-biographie.de/pnd118861476.html> [accessed 19 October 2023].

Scott, Tom, *Freiburg and the Breisgau: Town-Country Relations in the Age of Reformation and Peasants' War* (Oxford: Clarendon Press, 1986).

——— *Regional Identity and Economic Change: The Upper Rhine 1450–1600* (Oxford: Clarendon Press, 1997).

——— *The Early Reformation in Germany: Between Secular Impact and Radical Vision* (Farnham: Ashgate, 2013).

Stanford, Charlotte A., *Commemorating the Dead in Late Medieval Strasbourg: The Cathedral's Book of Donors and Its Use (1320–1521)* (Farnham: Ashgate, 2011).

Taylor, Larissa, *Soldiers of Christ: Preaching in Late Medieval and Reformation France* (Toronto: University of Toronto Press, 2002).

Tesauro, Magdala, Christine Hollenstein, Ramon Egli, Alain Geiger, and Hans-Gert Kahle, 'Continuous GPS and Broad-Scale Deformation across the Rhine Graben and the Alps,' *International Journal of Earth Sciences*, 94 (2005), 525–37.

Ustaszewski, Kamil, and Stefan M. Schmid, 'Control of Preexisting Faults on Geometry and Kinematics in the Northernmost Part of the Jura Fold-and-Thrust Belt,' *Tectonics*, 25 TC5003, doi: 10.1029/2005TC001915.

Voltmer, Rita, 'Political Preaching and a Design of Urban Reform: Johannes Geiler of Kayserberg and Strasbourg,' *Franciscan Studies*, 71 (2013), 71–87.

Wantzen, Karl M., Urs Ühlinger, Gerald Van der Velde, Rob S.E.W. Leuven, Laurent Schmitt, Jean-Nicolas Beisel, 'The Rhine River Basin,' in *Rivers of Europe*, eds. Klement Tockner, Christiane Zarfl, and Christopher T. Robinson, 2nd Edition (Amsterdam: Elsevier Ltd., 2022), 333–91.

Weiss, Paul, *1516–1700, heurs et malheurs d'une ville et d'une province: selon les "Chroniques" de Malachie Tschamser, père-gardien des Franciscains de Thann* (Boofzheim: ACM Édition, 2000).

Wetter, Oliver, and Christian Pfister, 'Spring-Summer Temperatures Reconstructed for Northern Switzerland and Southwestern Germany from Winter Rye Harvest Dates, 1454–1970,' *Climate of the Past*, 7 (2011), 1307–26.

Wieriks, Koos, and Anne Schulte-Wülwer-Leidig, 'Integrated Water Management for the Rhine River Basin, from Pollution Prevention to Ecosystem Improvement,' *Natural Resources Forum*, 21 (1997), 147–56.

Woods, Robert, 'Urban-Rural Mortality Differentials: An Unresolved Debate,' *Population and Development Review*, 29 (2003), 29–46.

2   Pre-reform Roman Catholicism and nature (Case Study 1)

Introduction

Vade ad formicam o piger et considera vias.[1]

With this quotation from Scripture as inspiration, Johannes Geiler von Kaysersberg stepped into the stone pulpit in Strasbourg's *Liebfrauenmünster* on 11 February 1509 (*21 February*) and started his annual cycle of 40 Lenten sermons plus preface; that year, his theme was the humble ant. A tiny denizen of the natural world, Geiler's understanding of the insect gave impetus to 16 sermons before he turned to another theme for the remaining 25: witches and unholy creatures. The value of these sermons for the first case study comes not from the moral lessons which were his primary purpose in preaching, but from the descriptions, anecdotes, analyses, and other references he provided about the natural world. That they were incidental and unexceptional to his audience emphasizes the widespread shared assumptions upon which they were based and provides a strong basis of comparison for imminent changes in the religious view of nature.

Die Emeis (The Ant) is not the only collection of Geiler's sermons to take the natural world for thematic inspiration. Sermons collected under the titles *Das buoch Arbore humana. Von dem menschlichen Baum* (The Human Tree Book. About the Human Tree; delivered during Lent for 1495 and 1496, and on into 1497), *Höllischer Löwe* (The Infernal Lion) and *Löwengeschrei* (Lions' Roar, both delivered in 1507), or *Das Irrig Schaff* (The Misguided Sheep, known to have been delivered in 1501, again in 1505 and 1508, and possibly at other times) all find Geiler preaching from nature's table. That is, in these sermons he took advantage of well-known elements of the environment (trees, lions, sheep) as the basis of his moral instructions. *Die Emeis*, however, was the last of this type of sermon, delivered by the preacher during his final year of life. It was published in 1516 and again in 1517, making the sermons present to the Strasbourg community as both lived memory and in print the same year as Luther was making his opposition to indulgences known in faraway Wittenberg. The timing of both Geiler's delivery of these sermons and their collection into *Die Emeis* gives

DOI: 10.4324/9781003262411-3

the opportunity to juxtapose them with early Lutheran and radical religious understandings of nature in Strasbourg, as the worldview presented by Geiler – and the ritual practices based upon it – would have been available to contemporaries when alternatives arrived with the preaching of Reformers Matthäus Zell, Martin Bucer, and Wolfgang Capito, and radical peasant theologians like Clemens Zyegler.[2]

Just as the previous chapter built a portrait of the society of early sixteenth-century Strasbourg and Alsace, this chapter describes the cultural and intellectual approach to nature against which later changes and events may be understood. Alsatian ideas about the natural world were not held independently, but in a living community where the ideas received from generations past became grounds for present action, both protective and remedial, and for later contestation. Accordingly, while this first case study and the two that follow are built around and emphasize a single unique religious representation of nature, it is crucial to remember that these were contested ideas expressed by people living alongside those whose views may have been very different.

After a brief biography and an outline of scholarship concerning Johannes Geiler, the chapter opens with a 36-year weather report interspersed with the economic repercussions of the environmental stresses of the period and some social responses within the context of the Alsatian social order outlined in Chapter 1. The following section, 'A reasonable man exorcises the storm: Orthodox methods of ameliorating extreme weather events in late medieval Alsace,' introduces apotropaic and salutary practices that were culturally available to pre-Trentian Roman Catholics for addressing extreme weather events, including magic, which, while not dogmatically endorsed, was accepted as genuine. Following Case Study 1, 'Natural philosophy, theology, and Geiler' explores the intellectual and theological foundations of Geiler's representations of nature. The chapter concludes by summarizing and exploring the concept of the natural world in Geiler's worldview.

Born on 16 March 1445 at Schaffhausen, Johannes Geiler moved to Ammerschweier, Alsace (near Kaysersberg), following the early death of his father. At 15 years of age, the young man entered the University of Freiburg's Faculty of Arts, earned his baccalaureate from that institution, and in 1464, his *magister artium*. From 1465 until 1468, he lectured in the Faculty of Arts there and in 1469 – at 24 years of age – was elected Dean. He occupied the post only a few months, though, as he left at the end of the semester in 1470. Enrolling as a doctoral student at the University of Basel during the summer of 1471, Geiler taught there at both the Faculties of Arts and Theology and was elected Dean in 1474. He was awarded a Doctorate in Theology in 1475. He was also ordained as a secular priest in Basel, serving in that city's cathedral by 1476 and likely offering his first sermons.

Geiler struck up a number of acquaintances while in Basel, including Wilhelm Textoris and Johannes Heynlin von Stein, both preachers at Basel's cathedral, as well as Johannes Reuchlin, Johannes Matthias von Gengenbach,

Johannes Ulrich Surgant, and Christoph von Utenheim, later Bishop of Basel; his close friendship with Sebastian Brant dates from those years. Nevertheless, Geiler returned to Freibourg-am-Breisgau in May 1476 to lecture and, the following year, to take up the position of Rector of the university. In Freibourg, he resumed a friendship with humanist and theologian Jakob Wimpfeling, likely first encountered during his earlier teaching post.[3]

Eventually to write one of Geiler's posthumous biographies, Wimpfeling reported that Geiler's own doubts about his ability to issue measured penance in the confession box led to his decision to focus on preaching as his life's vocation. His talents were quickly recognized, and the first city to offer him permanent and well-paid employment was Würzburg. Before Geiler could take up the offer, however, Peter Schott the Elder, sometime *Ammeister* (equivalent to Chief Magistrate) of Strasbourg and leader of the city's armed forces in the war against Burgundian Charles the Bold during the 1470s, persuaded him to accept the specially created position of preacher in Strasbourg's cathedral. On 1 April 1478, Ruprecht von Pflaz-Simmern, Bishop of Strasbourg, officially installed Johannes Geiler von Kaysersberg as the designated preacher in the *Liebfrauenmünster zu Straßburg.*

According to Beatus Rhenanus, Geiler's friend and, along with Wimpfeling, biographer, the preacher held to a very regular and disciplined personal schedule: awakening at 02h00 or 03h00 to compose the sermon he would deliver at 06h00, delivering the sermon and, upon return from church, writing down the sermon he had delivered; at 09h00 celebrating Mass, at 10h00 reciting the *Liturgia Horarum*, at 11h00 eating. At noon, he read while walking, at 13h00 he slept, and at 15h00 he researched the subject of his next sermon. After the evening's recitation of the Breviary, he allowed himself to have a walk or meet with friends.[4]

Geiler's agreement with his employers (the cathedral chapter) called for him to preach every Sunday and every day during Advent and Lent, as well as a certain number of feasts, and to join the processions required by the arrival of Legates and princes and collectively endured calamities. As well as duties centred on the cathedral, he was also to preach regularly in various other monasteries and convents in and around the city. He was empowered to announce the change of market days, if at short notice, as well as newly introduced indulgences, and was given four weeks of holiday *per annum*. For his services, Geiler was paid an annual salary of 200 florins at a time when the average salary was one florin per week. He occupied the post until his death on 10 March 1510, at almost 65 years of age.[5]

Regularly enjoying a remarkable circle of friends and associates, which, along with Rhenanus, Brant, and Wimpfeling, included Peter Schott, Gabriel Biel, and Jacob Sturm, Geiler was exposed to the main stream of contemporary intellectual activity and, particularly, the humanism associated with the northern Renaissance – although he was not himself a humanist. Theologically oriented towards the *via moderna* (see below), he was widely renowned as an unusually influential preacher both in his own day and afterwards.[6]

A prominent Catholic preacher on the eve of the Protestant Reformation, Geiler has been an attractive subject of scholarly inquiry. As Rita Voltmer pointed out in her authoritative 2005 study of Geiler, *Wie der Wächter auf dem Turm: ein Prediger und seine Stadt. Johannes Geiler von Kaysersberg (1445–1510) und Strassburg*, eulogies delivered by close friends Wimpfeling and Rhenanus have provided most of the information about Geiler's life. Wimpfeling focused on Geiler as 'the reform programme personified,' while Rhenanus presented him as an ideal humanist. These different interpretations were further complicated by Protestant Daniel Specklin later in the sixteenth century, when, through the falsification of dates, facts, and circumstances, along with the creation of a few false sermons, Specklin's *Straßburger Chronik* re-fashioned Geiler into a proto-Protestant. For centuries, historians mistakenly accepted Specklin's representation of the preacher as accurate, leading to strenuous and inconclusive debates featuring evidence from Wimpfeling, Rhenanus, and Specklin in an attempt to determine whether Geiler was a failed Catholic reformer, a northern Humanist, or a Protestant forerunner.[7]

Moving away from attempts to establish a narrow – and probably anachronistic – identity for Geiler, mid-twentieth-century historians began a wide-ranging exploration of other aspects of his work, including his relationship with his sources, his view of witchcraft, the XXI Articles he submitted to the Strasbourg Council in 1501, the rhetorical and stylistic aspects of his preaching, and, recently, his understanding of social norms and their reflections in insect allegories. In particular, Voltmer's impressive body of work details, among other things, the fundamental reforms of social programmes involved in Geiler's sincere ambition to 'instruct, correct, and cleanse the community entrusted to him.'[8]

While taking advantage of previous research on Geiler, this book is not focused on resolving questions about his role in the developing Protestant Reformation nor in expanding knowledge about his political and social influence. While note will be taken of the latter, a different concern is the focus of this investigation and it is one which may be fruitfully pursued alongside other analyses. That is, since Geiler was an influential participant in Strasbourgeois society of the early sixteenth century, his central role in transmitting the religious culture of his period renders these sermons into a rich resource for Catholic views of the natural world – views which, while we cannot be certain they were widely shared, were accepted as unimpeachably orthodox in their time.

The central historiographical purpose shaping this chapter and the previous is to provide a basis of comparison for the changes in cultural representations of the natural environment that were to follow. The institutions and relationships described herein were based on a shared worldview (excepting minority populations, such as Jewish communities), within which Roman Catholic concepts of nature explained events and justified individual and collective behaviour with respect to it. These concepts had developed during the long centuries of Alsatian Christianity and in the context of the Alsatian dialect, linguistically related to Alemannic Middle High German. The theological changes introduced during the Reformation and by more frequent use of the German language, as popularized by Martin Luther, are to be understood against these cultural norms.

Weather report from 1473 to 1509 and some social consequences of that weather

While the main cultural focus of this study is from 1509 to 1540, most of the individuals whose works serve as primary source material for this research project were born in the fifteenth century and would have endured the extreme weather of the last decades of the fifteenth century as well as that of the first decades of the sixteenth. To have a grasp of the role played by weather in social turbulence and cultural change, detailed knowledge of the weather conditions during this period is needed. Therefore, this description of weather events and the social responses of the people of Strasbourg and Alsace features an annual weather report that begins with the extreme conditions of 1473, concluding for now in winter 1509 with Geiler's sermons later collected as *Die Emeis*.

Much of the material in this section is based on primary and secondary sources, as discussed in Chapter 1. Other information is provided from surveys on the Tambora database, which is recognized as a collaborative climate and environmental history research environment. Contributions to the database come from a wide range of sources, are peer-reviewed, and are accessible by identifiable parameters such as temperature, precipitation, storms, floods, etc. with different regional, temporal, and thematic foci. As compared to other historical climatological databases such as ClimHist, Tambora provides the specific comment, as well as full source citations. This type of data is ideal for research such as this, since it was the annual variations and extreme events which framed and influenced social responses.[9]

It is also worth noting that many, if not most, weather-related records were kept by those whose livelihood relied on a successful harvest of wine grapes, which creates a grape-flavoured bias. However, the weather conditions for a successful year of grapes are also those which generally lead to a successful year for grains, vegetables, fruits, and beasts in the Upper Rhine Valley. Therefore, a harvest report for wine grapes is likely to reflect harvest results for other produce and unless salient differences, such as extreme heat and/or drought (when grapes may survive while other crops die), are specifically mentioned, will be understood as such here.

In 1473, notice of spring's early arrival was served when trees bloomed in February, grapevines bloomed in March and April, cherries were mature in March, and other fruit arrived in May. Grapes matured in June; crops ripened before St John's Day (*24 June/3 July*), and the grape harvest occurred in August – an event which usually took place in late September or October. Harvests of fruit and wine were good, but root and green vegetables were expensive, implying these crops were not as successful. The summer was hot and dry; beasts suffered and fires burned the forests of Mont Sainte-Odile (*S. Ottilienberg*) and the cloisters, with damage to the prestigious Abbey. The *Shwartz Wald* (Black Forest) on the east side of the Upper Rhine Valley also burned, as did forests farther away, like the *Böhmer Wald* (Bohemian Forest) and the *Thüringer Wald* (Thuringian Forest). The trees bloomed again in September and October.[10]

As a result of the extreme heat and aridity, winter grain harvesters employed by the Prince-Bishopric of Basel, for example, began their work on his properties on the earliest date of 517 recorded years studied. Perhaps the threat of fire descended from Mont Saint-Odile to the plains of the Upper Rhine Valley, because on Sunday, 8 August 1473 (*17 August*), just before the Feast of Saint Lawrence, patron saint against fire, Strasbourg's *Rat* commanded prayers for good weather in all the city's churches. The next day, bells were ordered to ring in all the monasteries, parishes, and cloisters at 05h00, followed by a small weather rogation (a communal supplication to God for mercy; see the next section for more information about rogations).[11]

Alsatians were not alone in their experience of this extreme weather event: a survey of the historical climatological database Tambora for the year 1473 suggests that the heat wave may have been a continental event, as exceptionally hot and dry seasons dominate the results. Less than 130 km to the northwest, for instance, on the other side of the Vosges Mountains, the citizens of Metz found cherries available for purchase on the ninth day of May, and a month later, local grapes were in *verjus* (soft enough to press for the sour acidic juice used in medieval cuisine). The high temperatures continued, with chronicler Jehan Aubrion describing many people as dying from the heat during the last ten days of July, and when a troop of soldiers went out from the city to harry enemies from Lorraine, he reports that the heat killed eight horses and many of the men returned sick. Further afield, chroniclers as far distant from Strasbourg as those in the German-speaking cities of Konstanz, Lindau (*Bodensee*), Ulm, Koblenz, Kitzingen, Eisfeld, Nuremburg, Munich, Regensburg, Halle, and Göttingen, as well as, in today's state of Poland, the Silesian towns of Świdnica (*Schweidnitz*), Strzegom (*Striegau*), and Legnica (*Liegnitz*), along with the town of Toruń, northeast of Warsaw, all mention abnormally hot temperatures, abnormally low levels of precipitation, and destructive fires. In the Elbe River near the northern Czech town of Děčín, the year 1473 was inscribed on a so-called 'hunger stone,' which, revealed during the drought of 2022, also reads *Wenn du mich siehst, dann weine* ('If you see me, then weep'). Yet even more distant from Strasbourg, the *Ustyug Letopis* from northwestern Russia also documents severe droughts and large fires for 1473.[12]

A couple of relatively less arduous years seem to have followed the extreme weather of 1473. There was relief from the drought and heat in Alsace in 1474, but on Saint Peter and Saint Paul's Day (*29 June/8 July*) that year, strong winds and storms were reported from Hungary to the Rhine and north to Lemwerder, close to the coast of the North Sea. In the Black Forest east of Strasbourg, they caused substantial property damage and on the Rhine River (exact location unknown), boats were overturned. Perhaps due to these weather events, wine grape harvests in the region were reported as being smaller than usual. The chronicles available for the Upper Rhine Valley hold no mention of 1475, leading one to hope that the weather may have been unexceptional and harvests successful.[13]

A very cold winter marked the beginning of 1476, however: the Rhine froze over, an event which had only occurred twice before during the fifteenth century. Spring was late and cool, the summer was cold, and harvests across the board were mediocre or poor. Similar conditions occurred again in 1477, when the growing season was reported to be cool and wet. People in Köln (*Cologne*) experienced a 'flood of the century' that year around Corpus Christi (*5/14 June*); although there is no mention made regarding water levels for the Upper Rhine, water levels may have been high in that location as well. In 1478, wine grape harvests were better than average in several regions north of the Alps, including the Upper Rhine Valley; although there is only silence regarding other agricultural products, there is also no mention of their failure – and while not reliable, silence implies unexceptionally average harvests. Favourable conditions continued in 1479; chronicles report a hot and dry summer, leading to some water shortages but in Alsace, to such an abundance of wine grapes and other produce that prices plummeted – an advantage to purchasers, but a disadvantage to producers.[14]

Destructive and extreme weather events returned in 1480, when the Rhine and other rivers north of the Alps, such as the Lahn and Moselle, flooded. *La Chronique Strasbourg* relates how rain fell steadily for nine weeks that summer and on the evening of Mary Magdalene's Day (*22/31 July*), sheaves of grain in the field were swept away by flooding waterways. The Rhine and the Ill Rivers were so swollen between Basel and Strasbourg, Meyer imparts, that none of the water mills remained erect and several people drowned. Many houses and villages were destroyed and some people had to cling to trees to survive. The water carried waste, worms, snakes, frogs, leeches, and other light vermin; the many deaths that followed were attributed to these having poisoned the air and the earth. Silbermann's notes on the *Collectanea* observe that Daniel Specklin mentioned the crisis, albeit with an emphasis on the city's fortifications: the architect had included a comment that 60 fathoms of the city walls fell down due to the flood. Within city walls, Strasbourg's *Rat* organized a rogation on 26 June (*5 July*) for 'peace, good weather, and all other needs' and a larger one on 5 August (*14 August*). Nevertheless, the harvest failed in Alsace that year, with reports of hunger among the poor.[15]

It failed again in 1481, thanks to another rainy and very cold summer; wine from the harvest in the Upper Rhine Valley was sour, the harvest was poor, and prices high. As noted by Francis Rapp, two years of such conditions created a fear of famine and made people desperate and angry. Months later, in the spring of 1482, the *Rat* acted apotropaically: just before the

Sunday preceding Rogation Week, they distributed a note to the preachers of all the city's churches. In it, they requested zealous urging of parishioners to join in the rogations begging God for good weather, in order that the harvest might succeed; a greater earnestness of Christian devotion was called upon, in hopes that God would provide succour. Indeed, supportive weather conditions that summer allowed for a very plentiful harvest of wine grapes in the Upper Rhine Valley, and likely other products. In 1483, a second year of hot and dry conditions contributed to an abundant harvest and low prices for wine in Alsace, reported in both Meyer's *La Chronique Strasbourgeoise* and Specklin's *Collectanea.*[16]

Floods and a visit of the plague arrived in Strasbourg and Alsace in 1484 – but a great deal of high-quality wine was produced along the Rhine for the third year in a row, so much that purchasing barrels to store it was more expensive than the wine was worth. Not all the grapes were harvested and, occasionally, water was forsaken for wine in the mixing of mortar. A full solar eclipse happened on the afternoon of Saint Gregory's Day (*12/21 March*) in 1485 and, that summer, another weather rogation was called by the *Rat*; harvests of that year were successful. However, the harvest of grapes was low in 1486, 1487, 1488, and 1489 for reasons not included in the archive; this led higher prices, implying below-average harvests not only of grapes but also other crops. Above-average costs continued into 1490, when a heavy snowfall thawed into spring floods at Strasbourg and Cologne. Many went hungry that year and came to Strasbourg, hoping for succour – which was given, even as plague wracked the city.[17]

Following these five years of low harvests, Alsatians and the rest of south-western Germany saw a very long, cold, and grim winter in 1491; the Rhine River froze again near Cologne, and with the thaw, flooded. Frosty temperatures and substantial snowfall in May froze grapevines and grain plants, as well as delaying the bloom of fruit trees. A cold and rainy summer followed, producing a small harvest of sour wine and very little fruit and causing the cost of grain to rise steeply throughout the southwestern region. Hunger among the poor was reported. Winter of 1492 continued cold, but the summer was hot and dry. Nevertheless, the wine harvest was scanty and sour; prices for farm products remained high and hunger was reported again. A meteorite landed in the fields near Ensisheim on 7 November (*16 November*) of that year; in hopes of miracles, pieces of it were mounted on church walls and kept in homes.[18]

These three years – 1491, 1492, and spring 1493 – saw extraordinarily high prices for grain and wine, a consequence of two years of weather-related critically poor harvests following five years of (likely) low harvests. For example, a quarter of wheat which cost 62 *Pfund-Pfenning* in 1487 cost 108 and 115 in 1491 and 1492, respectively, and 112 *Pfund-Pfenning* in 1493. To these heightened costs of living, add the quotas or rents owed by the peasants, the tithe – indexed on production – and a tax by landlords that fluctuated according to the needs of the moment (and, since Emperor Maximilian

I had been quartered in Alsace since November 1492, those needs were elevated). Peasants and poorer town-dwellers suffered, caught as they were between unfavourable environmental conditions and unsympathetic social superiors.[19]

On 23 March 1493 – the vernal equinox – around 35 Alsatian men climbed to the peak of the Ungersberg, a mountain of some 900 metres situated a few kilometres north of the silver mines at Saintes-Marie-aux-Mines in the Vosges mountain range southwest of Strasbourg. They came from ten different nearby localities owing allegiance to a variety of overlords, and they came to swear an oath to each other that every man and those he represented (totalling almost 1,500 *fussknechten*, or footsoldiers) would flame up in a violent demand for systemic change towards debt relief. Specifically, they demanded the abolishment of ecclesiastical courts (notorious for confiscating property pledged against a loan), a prohibition on the accumulation of ecclesiastical benefices, a ban on recourse to the Imperial Court at Rottweiler, and the expulsion of all the Jews from Alsace. Bischoff notes that the conspirators were likely not referring to the actual Imperial Court at Rottweiler, where arbitration was very lengthy and very expensive, but rather to the Vehmic courts, a secretive proto-vigilante tribunal system in Westphalia. Either way, their goal was to repatriate the seat of justice towards the litigants. To achieve their goals, the rebels were prepared to violently seize the town of Sélestat, where they had between 200 and 500 sympathizers, and then spread the revolutionary fire to the rest of the valley on both sides of the Rhine.[20]

Within ten days of scaling the Ungersberg, the plan of the conspirators was exposed. Authorities, among them the Bishop of Strasbourg and the *Rat* of the nearby Imperial Cities of Obernai and Sélestat, arrested the oath-takers and punished most of the hopeful with hefty fines and, unless extra sums were paid, the loss of each man's thumb and first three fingers of the right hand. However, a prominent leader in the plot was Hans Ulmann, one-time military commander, respected cattle dealer, and, as holder of civic office in Sélestant, host to Emperor Maximilian I five months before he climbed the Ungersberg. His disloyalty was perceived as informed and particularly heinous: arrested during an attempt to escape to Basel, Ulmann was drawn and quartered as a powerful caution to others with his status.

The Ungersberg plot – which failed to manifest into action – led, nevertheless, to two enduring legacies. First, the pact into which the peasants had entered characterized a new level of organization and resistance to landowners in Alsace; it became referenced from 4 April 1493 as the *Bundschuh* rebellion ('rebellion of the alliance of bound shoes,' pieces of leather tied up with laces commonly worn by peasants as footwear). Second, Maximilian personally requested the reorganization of the village courts, which doubled as municipal councils, of those villages from which conspirators had come. From henceforth, litigants would be directly questioned by an imperial provost and appeals would be remanded to Obernai; those who sat on the court would be considered as under imperial authority, their seats would be renewed every year, and the court would sit every Monday. Other provisions were aimed at the amelioration of economic problems, such as better control of the real estate market or the prevention of similar gatherings in the future.[21]

Indifferent to social tension, a late frost in April 1494 withered vines in the Breisgau area and, presumably, other young grain crops; the cold snap spread all the way to the Baltic Sea, where the frozen ocean between Sweden and Germany could bear the weight of a man. This was followed by a hot and very dry summer, leading once again to low harvests. Prices remained high, Alsace descended into famine, and some died.[22]

Finally, in 1495, weather conditions allowed good harvests to return: the region enjoyed a very good grape harvest, such that the harvesters could not pick them off the vine quickly enough, and all prices dropped to affordable levels, implying that the harvest of grain and produce was good as well. The Strasbourg *Rat* felt confident enough in the city's resources to build the *Blatterhauß*, a hospice for sufferers of the syphilis bacteria. There were average harvests throughout the Upper Rhine Valley in 1496, despite a severe flood by the Rhine River. While the high waters washed away a tower in the city wall and caused damage to the cathedral, Strasbourg's bridge across the Rhine survived. Although a wet and warm spring in 1497 produced an abundance of wine, grain prices in Alsace rose – implying local shortages. The next year, 1498, saw a low wine harvest in the Upper Rhine Valley, but southwestern Germany, as a region, had somewhat better-than-average harvests of wine in 1499. The Swabian War spilled over into southern Alsace that year, though, where experienced Swiss soldiers confirmed their independence from Maximilian I and the dukes of Burgundy by defeating the peasant-filled Swabian and Habsburg armies.[23]

The turn of the calendar to 1500 did not usher in good weather conditions, although the apocalypse threatened by astrologers and preached by Johannes Geiler failed to manifest. Tschamser's *Annales* mention that there was widespread harvest failure that year due to cold temperatures; other sources note that on 29 May (*8 June*), icicles were reported to be hanging from the grapevines in the Breisgau, an area between the Rhine River and the foothills of the Black Forest. This, along with excessive rain and hail, led to a rise in grain prices in 1500. In 1501, particularly cold temperatures were reported in August and, according to a chronicler in the convent at Thann, just east of Mulhouse, this was a year of need and tragedy as famine ruled in Alsace and in Germany.[24]

After a painfully cold winter in the Upper Rhine Valley, one so severe that birds froze or starved to death, on 3 April 1502 (*13 April*) a tired pedestrian arrived at the Udenheim residence of the Bishop of Speyer and told the prelate's butler everything he knew about plans for another peasant rebellion. Centred on the wine-growing village of Untergrombach, on the eastern bank of the Rhine River almost 90 km northeast of Strasbourg, the conspirators intended to seize Bruchsal and the neighbouring villages, then the towns in the area, the episcopal castle, and the abbey of Maulbronn. Led by local

man Joss Fritz and timed for 23 April (3 May), the uprising was reputed to
have 400 eager rebels waiting to assist Fritz directly and an estimated 20,000
supporters around the region awaiting the word to join. Their goal was to
find freedom 'in the Swiss style' and an emissary had been dispatched to
the winners of the Swabian War. The conspiracy collapsed once the traitor
had revealed their plans, but Fritz eluded capture. Extracted under torture
and correspondingly unreliable, the list of demands included the abolition of
serfdom and all ground rents, tithes, taxes and tolls, as well as the division
of monasteries and church property among the peasants. In their refusal to
accept the territorial authority of any landowner except the Emperor, the Untergrombach rebels apparently attempted to circumvent most of the Alsatian
landowning class of the area.[25]

This second iteration of peasant willingness to conspire against the status
quo awoke frightening memories of the plot of the previous decade and Fritz'
scheme was quickly called a *Bundshuh* rebellion. The alarm of those whose
position and wealth were threatened, the Alsatian landowners, inspired a
gathering in Sélestat on 29 April (9 May); the Holy Roman Emperor was formally represented by Louis de Mascvaux, lieutenant of the provincial government of Ensisheim, the bishop by his chancellor and by the gentleman Hans
von Mittelhausen, the count palatine by officers from the imperial prefecture
of Haguenau, and the counts of Bitche and Hanau by two other nobles. Guillaume de Ribeaupierre, administrator of imperial possessions on the Rhine,
companion to Maximilian I, and possessor of the hunting rights for all Alsace,
came in person; representatives came from the free imperial city of Strasbourg
and nine of the ten cities of the Decapolis (Mulhouse excused itself).

The ten-point plan developed by this conspiration of the elite focused on
speedy communication of information and the active prevention of further
peasant alliances. In the event of future reports of the *Bundschuh*, relevant
early information would be sent to the emperor and all those gathered, as
well as those nobles not present (the Duke of Württemberg for the seigneuries
of Riquewihr and Horbourg, the Count of Lupfen for Haupt-Landsbourg,
and the 'king of Sicily' – the duke of Lorraine – for Saint Hippolyte). Near
neighbours agreed to support each other in suppressing the unruly and four
'risk zones' were identified in which reinforced solidarity would play out: if an
insurrection were to develop south of the Landgraben, in Sundgau and in the
north of the Jura, the initiative to alert the others would be taken by the Austrian bailiff of Ensisheim and then relayed by Colmar in the direction of the
bishop of Strasbourg, who would transmit it to Lichtenberg and to the imperial prefecture of Haguenau, itself in charge of Wissembourg. The second zone
was centred on Colmar, the third on a Sélestat-Strasbourg axis and the last on
Haguenau. This plan covered the entire Upper Rhine Valley, from Wissembourg in the north to Basel (included but not present at Sélestat) in the south.[26]

Preventative measures included tighter controls on the inventory of weapons, ammunition, and, in case of a siege or distant operations, provisions. In
the event, tighter controls would be exerted also on the movement of travellers,

particularly men-at-arms, with a prohibition on the accommodation of unemployed *landsknecht* and vagabonds of any kind. Those who needed to travel would be obliged to seek express authorization, and innkeepers would find themselves prohibited from serving wine or allowing card and dice games after curfew, fixed at 21h00 in the villages belonging to Strasbourg's cathedral chapter. Then locals would have 15 minutes to clear the premises and guests would be sent to their rooms. This coalition and these measures would prove vital to maintaining elite authority over the next 23 years.[27]

As well as laying the groundwork for repressing peasant uprisings, some participating landowners recognized the economic challenges facing peasants. Guillaume de Ribeaupierre, for example, owner of silver mines at Ste-Marie-aux-Mines in the Lièpvrette Valley of the Vosges Mountains, negotiated investment from Sigismund of Austria which, later in 1503, allowed mining operations to resume. By the mid-sixteenth century, mines in the valley may have employed up to 10,000 men, with a potential supporting population of 50,000 – larger than the population of Strasbourg at the time.[28]

Weather in 1503 proved as unfriendly to peasants as the landowners had the previous year: another painfully cold winter, followed by a summer featuring excessively high temperatures and a drought endured from 16 May (*26 May*) until 2 July (*12 July*) and in places until 10 August (*20 August*). Such hot and dry conditions stimulate the growth of grapes, but while the wine harvest was abundant, it was of poor quality and the weather was disastrous for wheat, rye, and barley – the caloric foundations of the sixteenth-century Alsatian diet. Domestic animals also suffered and died, particularly pigs. The utter absence of rainfall led to a summer famine in Alsace. On 21 June 1503 (*1 July*), Albrecht of Bavaria, Bishop of Strasbourg, ordered weather rogations and extra prayers throughout the region. Tschamser's *Annales* includes mentions of several other days of repentance and fasting, as well as processions and rogations, in efforts to bring rain and avert further evils.[29]

Similar conditions prevailed for the next four years, not only in the Upper Rhine Valley but throughout southwestern Germany. In summer 1504, there was extreme heat, drought, wildfires, and a very good wine harvest in Alsace, leading to lower prices of that product. During the winter, although generally mild temperatures prevailed, snow covered the ground for most of February; a quick melt in March 1505 led to floods, but a warm and dry summer produced a very abundant wine harvest – so much so that old wine was poured out of storage containers to make room for the new. Unfortunately, the second summer of aridity led to the deaths of many beasts, as, along with a few humans, they perished from thirst. In 1506, a comet shone in the night skies and despite significant damage caused by a hailstorm on the Sunday after St. Bartholomew's Day (*24 August/3 September*), most harvests of wine, produce, and grains were abundant and successful.[30]

However, in 1507, excessively high temperatures and severe storms such as those on 9 June (*19 June*) and the Monday before Saint Bartholomew's Day (*24 August/3 September*) once again caused dearth in the bioregion.

A storm reported on the night of Saint Peter and Paul's Day (*29 June/9 July*) brought winds to the Upper Rhine Valley so powerful that trees were uprooted and thrown, houses scoured of roof tiles, and buildings were shaken. Hail followed the wind, damaging fruit trees and wine grapes in the Breisgau. Heat lowered the water table and prices rose for bread and vegetables, but the wine harvest was good. Cooler temperatures and enough rain to restore the water table arrived in 1508, but 'enough' rain turned into continual rain during the summer, which weakened an already-enfeebled swine population; disease depleted the pigs of the region. A warm autumn, however, rescued the harvest and prices were affordable. Following a very snowy winter, Johannes Geiler stepped into the pulpit carved specially for him in Strasbourg's *Liebfrauenmünster* and delivered his sermons for Lent of 1509.[31]

This section provides the first 36 years of the volume's 68-year weather report, from 1473 to 1509, and some of the social consequences of that weather. To summarize: during these years, 20 harvests were reported to be above average (1479, 1482, 1483, 1484, 1495, 1499, 1505, and 1506) and average (1473, 1474, 1475, 1478, 1485, 1493, 1496, 1497, 1498, 1502, 1504, 1508); the absence of any mention of an annual harvest in the archives implies that they were unexceptional, unremarkable, and, therefore, considered here to have been average. During the same period, 12 harvests were reported to be poor (1476, 1477, 1480, 1481, 1486, 1487, 1488, 1489, 1490, 1491, 1492, and 1507) and four harvests failed completely (1494, 1500, 1501, 1503).

The poor of Alsace, whether rural or urban, endured dearth (scarce food supply) six times during this period: in 1480 and 1481, following the poor harvests of those two years; a decade later in 1490, 1491, and 1492, again following poor harvests; and in 1507, after the last poor harvest of this section of the weather report. The entire society suffered famine (drastic food shortage) four times from 1473 to 1509: in 1494, 1500, 1501, and 1503. Those famines of 1494, 1501, and 1503 are explicitly ascribed to harvest failure in the historical archive; although the harvest failed in 1500 as well, distribution of saved resources must have been sufficient to avoid regional famine.

Two sequences of bad weather contributed to peasant plans for achieving an easement of economic pressure from social superiors: in March 1493, the spring immediately following the long string of poor harvests from 1486 to 1492, and in April 1502, the spring immediately following the harvest failures of 1500 and 1501. Their explicit goals were focused on debt relief; the methods advised by the Roman Catholic Church did not succeed in providing enough relief from environmental conditions.

A reasonable man exorcises the storm: Orthodox methods of ameliorating extreme weather events in late medieval Alsace

Weather is never meaningless. In the twenty-first century, weather contributes to, among other things, the accumulation of climatic data, to conversation, and, as in the sixteenth century, to economic stability and the continued well-being of creatures who experience it. The most abstract theme of this book is the great transformation in commonly held views of nature introduced by the Protestant Reform; in order to understand the scale and scope of that change, a basis of comparison from late medieval Roman Catholic orthodoxy is useful. When Johannes Geiler von Kaysersberg delivered his first Lenten sermon in 1509, he addressed a community that was, if not united, at least agreed on the basic structure of the natural environment, how it was supposed to behave, and the range of possible reasons for unusual events. The extent to which Geiler's comments about nature, as well as the philosophical and theological arguments upon which they were based, were accepted as axiomatic truths by other people in the society wherein he preached can be demonstrated by considerable popular willingness to participate in ritual behaviour legitimated by the views he expressed.

Some caution is merited in fully accepting the behaviour of the early sixteenth-century Alsatians as proof of their unquestioning adherence to views presented from the pulpit, as sincere faith cannot be confirmed by behaviour from which withdrawal could earn social or political censure. Nonetheless, evidence suggests that other members of Geiler's community, whether urban, rural, literate, oral, exalted, or reviled, made the same intellectual and spiritual assumptions regarding the natural world that he did, and, for the most part, were prepared to act upon them. Bell-ringing, sacred architecture, communal rituals, and exorcisms were demanded by local people and included most as participants, demonstrating close cooperation between civil and religious authorities in addressing the multi-faceted threat of bad weather. Each of these prophylactic and curative measures will be described below.

Bells were familiar and well-known in Alsace and since they were a common element of public space during the medieval period, both clergy and laity were fluent in the language of their tolling. Distinctions between the bells and the nature of their peals indicated to which manner of event the listeners were summoned, whether of religious or civil import. Clangs could summon those within listening distance to a distribution of alms, the celebration of a particular feast, or warn of a military attack upon the community, as well as indicate the time or protect against storms. The sound of a larger bell ringing in a higher tower would carry further than its smaller, lower competitor, rendering the size of the bell and the height of its tower representative of the authority of its ringer. Unsurprisingly, bells and towers were tightly regulated for size and height.[32]

The *Liebfrauenmünster zu Straßburg* enjoyed an array of bells, with at least five different ones hanging in the cathedral by 1500. In 1411, the chapter cast and installed a bell to be rung specifically against bad weather; Sebastian Brant notes in his *Annales* that the bell was re-baptized and awakened again for duty in 1451. Alfred Pfleger's research reports that common people throughout Alsace referred to church bells as the 'Hounds of God,' whose 'barking' was unendurable to the devil and any weather-making witches within hearing. Among others, the cloisters in Murbach had a bell known as the Big Dog, Bühl bei Gebweiler, near Mulhouse, had St. John's

Dog hanging in its tower, Mühlbach im Münstertal had Bartholomew's Dog to ring out, and in Kaysersberg, close to Geiler's childhood home, the *Oberhofhund*, or Superior Court Dog. These were all explicitly known as weather bells, to be rung as a way of dispersing storms and other threatening weather events. We know that they were used regularly, as documents mention bells being rung against bad weather throughout Alsace, such as in Dossenheim (Dossenheim-Kochersberg, 1481), Börsch (1487), Schlettstadt (Sélestat, 1498), and Hunaweier (beginning of the sixteenth century).[33]

Other aspects of church architecture also provided protection against destructive weather, albeit more passively and locally; specifically, the sanctity of the Word was used to protect buildings. On the upper part of the dome in Strasbourg's cathedral under the chandelier, for example, there was an inscription on the western side: 'God protect me henceforth from thunder, hail, storms.' On the upper galleries of the same building, engravings of four sayings from the beginning of the Gospel of John (John 1:14) were made, aligned to the four cardinal directions and capped with the abbreviation of the name of Jesus Christ. The Gospel of John was associated closely with protection against foul weather, and, while there is no direct confirmation of it, it is likely that these inscriptions were carved with the goal of protecting the cathedral against lightning and other elemental assaults.[34]

Demanding a much higher degree of organization than ringing bells or making engravings were the ritual observances known as rogations, that is, communal processions to beseech God's protection. An early Christian development from Roman processions pleading with the goddess Robigo for her protection of crops, these regular and common ritual observances in late medieval Europe were conducted not only to plead for security from bad weather but also for protection from other threats such as plague or invading armies. Drawing upon Lawrence M. Bryant's perception of processions as public spectacles which give participants the opportunity to perform their religious and political identities, the composition of rogations and the route they followed was anything but casual. What follows is an exploration of those details in Strasbourg, a few words concerning the frequency of the ceremony, then some information about the route of weather rogations through the city.[35]

During the twelfth century, Sicardus of Cremona (1155–1215) included a description of rogations in his book *Mitrale* which was copied by Guillaume Durand at the end of the thirteenth century in his *Rationale divinorum officiorum*. The popularity of the latter volume, which survives in several instances, spread Sicardus' version throughout western Europe. The order of participants in the procession was prescribed as follows: positions at the beginning would be occupied by the clergy, followed by the religious male orders, nuns, novices, laymen, widows, and, at the tail end, married women. In his version, Durand added that the crucifix and reliquaries should be carried at the very front, so that the war banner of the cross and the prayers of the saints might clear away any demons who would face them.[36]

Strasbourg's rogations did not deviate in any significant way from the model established by Sicardus and Durand. The entire religious community was required to produce a successful rogation, which, by extension, meant that the organizers of the ritual demanded the participation of the great majority of those dwelling within city walls (clergy, canons, monks, parish clergy, and their parishioners). All the workmen and artisans were required to participate through their guilds and to ensure that upper-ranking citizens did the same, the *Rat* legislated on three consecutive years (1469, 1470, and 1471) that no person should depart the town and suburbs on Wednesdays and Thursdays without special permission from the Ammeister.

Johan Twinger von Königshoven described a weather rogation in Strasbourg in his *Chronik* of the early century. He notes that in the 1401, eight days after Candlemas (*2/12 February*), a rogation was organized in Strasbourg for the sake of good weather. It had been inspired by such heavy and on-going rains that daily life was difficult, made more so by a dearth of grain and fruit in the region. All the charities, the lame, and the cloistered were to begin at their own churches with the sacrament, and join together at the cathedral, where the rogation would be organized. Two students carrying flags went first, followed by vicars and canons. After them came Franciscans monks, carrying the holy cross which regularly was positioned behind the front altar of the cathedral; they were followed by those carrying the sacrament with candles and little bells. Then came the men of the community, followed by preachers carrying an image of the Blessed Virgin Mary; the women of the community followed her. Any Strasbourgeois thinking they might watch the rogation from their window or door were discouraged from doing by threat of a fine.[37]

During the decades which followed Twinger's death, the precise position any individual occupied in Strasbourg's rogations became so contentious that in 1472, the *Rat* legislated the composition and choreography of the Corpus Christi rogation. Little had changed from Twinger's time, although careful attention was given to the placement and responsibilities of city officials; these worthies were to gather at the cathedral's rood screen in preparation, they were to walk immediately following the holy sacrament and before the guild candles, and their servants were to assist throughout.[38]

Other than for the purposes of beseeching good weather, rogations were a regular element in the liturgical year and are known to have taken place in Strasbourg on St. Adolf's Day (*11/21 February*), Pentecost, Corpus Christi, Saint Luke's Day (*18/28 October*), and others. Among the regular cycle, entreating God for good weather leading to a successful harvest was the particular focus of the major litany held on Saint Mark's Day (*25 April/5 May*) and the minor litany held on three annual Rogation Days held immediately before Ascension (*15/25 August*). Outside of regularly scheduled rogations, Strasbourg civil authorities organized the ritual in response to unusual or unexpected misfortunes, such as extreme weather events or earthquakes, or against military threat, such as occurred during the Burgundian Wars.[39]

Two types of rogation were conducted in Strasbourg: smaller and larger. Small processions were identified thusly because they were limited to symbolic perimeters of various parish, monastic, and collegiate churches, and excluded the *Hochstift* (the territory of secular authority held by the Bishop of Strasbourg). The greater rogations solemnly circled the entire old town and sprang from the joint initiative of the Bishop, secular clergy, religious orders, and city council. The scheduled annual weather rogations were large events, as were many of the specially organized weather processions, which emphasizes the importance attributed to them by organizers and participants.[40]

The traditional path followed by large rogations was as follows: after having gathered around 08h00, participants departed through the main door of the cathedral and went straight ahead, past the hospital on what is now Rue Mercier, to pause for the first prayers at the parish church of Saint Martin. Crossing Rue des Ferruriers, they paused again for prayer at Saint Thomas' church, then went down the Grande Rue until the stone bridge by the tanners' ditch (possibly Pont des Faisans) and stopped to pray. From there they passed before L'Église de Saint Pierre-le-Vieux, carried on to the old wine market building, where a fourth prayer pause was held, and on to the *kirchgass* (possibly Rue du Dôme today) for a fifth prayer pause. The procession wound down Rue de la Nuée Bleue until the horse market and the Gürtlerhof, where the sixth pause for prayer occurred. Crossing the Rue des Prêtres, they halted at the wells of Saint Laurent's chapel and prayed; turning behind the Bruderhoff, they went to the Bishop's palace, where the eighth and final prayers were delivered. From there, the procession returned to the nearby cathedral and entered through the main doors.[41]

Occasionally lasting for an entire day, the procession criss-crossed the island of Strasbourg and nodded to all four cardinal directions around the Münster. Direct evidence of Geiler's participation in rogations of any kind is lacking. However, the demand by the *Rat* for full participation from all secular and regular priests in the city combines with the stipulations in Geiler's contract that he preach for processions required by public calamities to establish convincing circumstantial evidence for his presence as a reliable member of the clerical component of Strasbourg's weather rogations. Among the earliest rogations with which Geiler might have been involved as preacher and participant was likely to have been held on 26 June 1480 for 'peace, good weather, and all other needs'.[42]

Rogations pleading for God's protection were not limited to the urban centre, but were valued by communities throughout Alsace. For example, in 1510, as part of their argument to the city's *Rat* for ecclesiastical independence from the cathedral chapter, the commune of Ostwald included their need for the protective powers of the rogation ceremony in their list of reasons for greater autonomy. A village southwest of the city on the Ill River, Ostwald had been annexed by Strasbourg in 1418, which incorporated the small community into the city's administrative orbit and made the appointment of a priest to the local church the prerogative of the cathedral chapter.

In the third item of the document pleading their case for a local appointee, the village officials articulated their dismay about the absence of rogations to protect the fruit against thunder, hail, snow, flood, and other bad weather, and the need for a local priest to lead them.[43]

Should the attempt to shield fields, orchards, or crops from storms and other destructive weather events by rogation fail, one final measure remained to be taken: direct blessing of approaching storm clouds by an ordained priest in order to exorcise the weather demons. These instructions from a fifteenth-century document held in the Strasbourg State Archive are particularly vivid: not in a shelter, but on the land where the blessing is desired, the celebrant is to begin by placing his right foot on the earth pointed against the storm and with his right thumb, under his foot, mark three crosses on the earth in the name of the Father, the Son, and the Holy Ghost. After this, he should stand up and with his right hand, strike a blow against the storm and commend it to almighty God, his dear Mother, and all the saints.

Once the priest had done his part, the anonymous author included a vision of the spiritual response which would follow: the Blessed Virgin Mary would carry the request to her son Jesus Christ, who sits in a corridor on a golden stool. Christ's response would be that the saints Luke, Mark, Matthew, and John were to command in his name that no bad weather should strike or burn anyone who called on the Holy Spirit for help. He would also promise to cause the thunder and hail to release themselves upon the earth with as little damage as his mother had borne under her chaste heart, in God's name. Meanwhile, back on earth, the priest and his community should pray 15 Our Fathers, 15 Ave Marias, and, for Saint Pelagius' sake, distribute two bushels of bread as alms on the Friday of Easter Week. If hailstones were seen to fall, then after offering the blessing, a small handful of the hailstones should be collected and tossed in a fire.[44]

The Catholic conventions of engraved protection, apotropaic ceremonies, and direct blessings when destructive events were imminent were the primary means available to guard Alsatians against extreme weather events. Such means were frequently employed. Although, once again, it is important to avoid confusing social practice with heart-felt faith, the widespread demand for and communal participation in such events offers convincing evidence that the ceremonies were commonly accepted as reliable means of bringing the natural world into alignment with human needs. Since these methods were based on the same theological and natural philosophical view which were the foundation of Geiler's 1509 Lenten sermons, it stands to reason that Geiler's understanding and representations of the natural environment were common for his time.

Case Study 1: The natural environment in *Die Emeis*

Given the Catholic worldview held by late medieval Alsatians and the social dominion of the Roman Catholic Church, we can look to the words and behaviour of orthodox religious leaders for an understanding of the

Catholic concept of the natural world at the beginning of the sixteenth century. A note of caution: their words and behaviour cannot be considered an official policy response of the Roman Catholic corporate body to material evidence of climate change, as neither the concept of climate nor the comparison of long-term quantitative measurement were available at that time, and the reach of late medieval Christendom included several different climatic zones. However, the sermons of respected orthodox leaders such as Johannes Geiler provide insight at least into local and regional attitudes towards nature which were acceptable to both the clerical hierarchy and the wide range of parishioners. This section of the chapter begins by noting a curious absence in Geiler's sermons of references to the specific weather conditions he experienced or the challenges they posed for his parishioners, and briefly explores a sermon exciting particular expectations for such references and its source material. The bulk of the section is devoted to reporting the manner in which Geiler represented the natural environment in his Lenten sermons of 1509, collected as *Die Emeis* (The Ant).[45]

A word about the approach taken here to the subject of Geiler's views on witchcraft and magic: for most late medieval Europeans, magic was understood to be a tool or an instrument which allowed individuals to assume control over many features of daily life, including the natural world, and Geiler's views on the subject have occupied the attentions of several scholars. Early research approached the issue as a moral question, either laying blame for a belief in sorcery at the feet of Protestantism or as part of a defence against this accusation. More recently, the focus of the research has largely centred around the question of Geiler's personal belief in the material reality of magic, as encapsulated in the issue of flight by witches. Did Geiler think that the claim of flight by witches was a diabolical delusion of the witch? Or did he believe that witches could truly take flight? It was considered that resolving the issue would reveal whether Geiler upheld the *Canon Episcopi* from the twelfth century or was persuaded by fifteenth-century authors such as Heinrich Kramer or Ulrich Molitor that witchcraft was a clear and present danger. In her 2013 and 2017 discussions of Geiler's view of magic, Rita Voltmer found that Geiler – along with many other scholastically trained theologians of the period – presented ambivalent, even contradictory views about this matter. Over his lifetime, his sermons intermingle ideas that flight was both a diabolically implanted illusion and that it was materially possible, if only feasibly accomplished by the devil. Voltmer concludes that, while he may not have accepted unquestioningly the ideas presented in the *Malleus Maleficarium*, Geiler's preaching on the subject mediated those ideas for his audience, re-interpreting behaviour which had been hitherto innocent (if not entirely reputable) as sinful, criminal, and potentially deadly.[46]

The topic of witchcraft, magic, and the supernatural in *Die Emeis* will be explored in greater detail in the next section, but a caution is advisable with regard to the focus which will be pursued during this case study: while scholars of late medieval witchcraft and magic may examine *Die Emeis* for

Geiler's views on the subject, this volume is interested in his views on the natural world. Since magic was understood as a useful tool to manipulate the natural environment, it is included in this research only for what it reveals about Geiler's view of nature.

Unlike forward-looking Advent, which anticipates the celebration of Christ's birth with the return of the sun after Winter Solstice, Lent asks the Christian to look back in self-assessment and, in commemoration of Christ's sacrifice of himself for the redemption of humanity, offers to the faithful an opportunity for repentance, penitence, charity, and self-abnegation. The dates of Lent are established by the date of Easter, itself a movable feast held on the first Sunday after the full moon following the Vernal Equinox. Lent falls at the tail end of winter in Alsace, usually before spring greens have begun to grow or the winter wheat ripens. During a ritualized period of the year featuring an emphasis on introspection and a review of past events and behaviour, when bodily survival was reliant upon a successful harvest from the previous summer, a twenty-first-century reader expects Geiler to have addressed events from the preceding months during his 1509 Lenten sermons – particularly since weather conditions in 1507 and 1508 had made the Alsatian harvest and, consequently, the prosperity or survival of Alsatian individuals and communities an uncertain question.

However, there was no mention in these sermons of current weather conditions or weather events of the recent past. The themes which were addressed provided ample opportunity for the subject, such as on the Monday following Reminiscere, the second Sunday of Lent, when Geiler's fourteenth sermon of the cycle was focused on the ants' ability to foretell the weather. He considered this ability to be prophecy, which is shown by the ants remaining in their houses when it is about to rain or snow. From this, he instructed his listeners to learn that just as the ants warned them of the coming rain, so there are seven prophets; three of them are good and four are evil, who will lie to whoever attends them. Later in the sermon Geiler drew a very brief analogy between the prophetic ant and a Godly preacher, crediting both (and perhaps himself) with prognosticative powers. Other than that, he followed the usual pattern of a thematic sermon interpreted according to the *Quadriga*, and turned from his knowledge of ants' behaviour (the literal sense) to an instructive moral discourse about the trustworthiness of various types of prophets – despite the obvious utility of a weather-forecasting insect for the regular challenges faced by the people listening to him.[47]

For inspiration on the content of his Lenten sermons, we know that Geiler drew from a variety of material: two works by Johannes Nider (two collections of sermons entitled *Formicarius* and *Praeceptorium divinae legis*) and a commentary on the Ten Commandments by the same author, as well as the *Opusculum de sagis maleficis* by Tübingen priest Martin Plantsch and the *Malleus Maleficarum*. This infamous last work was authored by Alsatian Heinrich Kramer, who was born in Sélestat c. 1430 and later styled himself as Henricus Institutoris. Jacobus Sprenger, who died in Strasbourg in 1495, is frequently associated with the work, but it is now thought that his association was more a result of Kramer's desire to bring credibility to it than active contribution from Sprenger. Geiler primarily used *Formicarius* as a template for 'Die Emeis,' albeit with great freedom; the same latitude is found in 'Von den Hexen und Unholden,' where the preacher plumbed the *Praeceptorium divinae legis*, the *Malleus Maleficarum*, and the *Opusculum de sagis maleficis* for his material. The *Malleus Maleficarum* itself incorporated much material from Nider's *Praeceptorium divinae legis*, occasionally directly, as well as from Book V of *Formicarius*. However, Gerhard Bauer's careful inter-textual analysis of Geiler's twenty-second sermon (part of 'Von den Hexen und Un-holden') shows that the preacher used all four volumes as source material, as passages in the sermon are found in the *Malleus Maleficarum*, but not in *Praeceptorium divinae legis*, and vice versa. Nider's works, then, emerge as the primary sources for Geiler's *Die Emeis*, with important secondary sources being the *Opusculum de sagis maleficis* and the *Malleus Maleficarum*.[48]

The direct model for Geiler's fourteenth sermon was the second chapter of the second book in Nider's *Formicarius*, where, in comparison to Geiler, the Dominican's reference to the literal behaviour of ants was cursory. Nider represented them as prognosticators and attributed their ability directly to divine instinct and, as Geiler would follow, Nider went on to address a sub-ject entirely unrelated to the weather: the role of faith in prophetic ability. The humble ant did not make another appearance until the onset of his next chapter. If Geiler did not speak about current or recent weather events, even with the opportunity provided by the interpretation of Scripture in a sermon framed by weather-forecasting ants and despite recent events which clearly did not conform to the hopes or even needs of the community, what *did* he have to say about the natural environment?[49]

Most vividly, the natural world in *Die Emeis* is the source domain for the many metaphors Geiler constructs about the human soul and Alsatian society. One such example is found in his second sermon, delivered on Ash Wednesday, when Geiler told his audience that the smallness of the ant cor-responds to humility and their black colour to simplicity; he went on to build his sermon around examples of humble and simple Christians beloved by God. During the next day's sermon (which we will visit again below), he asserted that an ostrich hen, who lays her eggs in warm sand but does not brood them, was the Alsatian ruling class, who abandoned their eggs (pious subjects) and failed to brood them into honourable and spiritual souls.[50]

The natural world's function as the source domain for metaphors of this type is found frequently in *Die Emeis*, particularly in the first section, 'Die Emeis.' However good-humoured or pointed such metaphors may be, they reveal to only a limited degree the meaning nature held for Geiler, due to the way metaphors are constructed. The source domain (in this case, the ant) exerts a formative influence over the metaphor, as the qualities available for metaphoric construction are limited to the qualities of ants. For example, Geiler would have been unlikely to construct a metaphor involving flight, as ants generally lack that capacity, and there is, indeed, little discussion of topics such as human souls soaring or flying to God's side in *Die Emeis*. Metaphorical meaning is found in the associations chosen for the target domain; Geiler's choice of the ant as the source domain for his sermons speaks more about his view of the target domain – whether the human soul, the body of the Church, or the quality of prophets, among others – than his desire to preach about ants.

A more direct manner of discovering Geiler's view of nature is available from his sermons, for, while his primary purpose for preaching was the salvation of human souls, one of many avenues towards achieving that goal was through instruction in orthodox cosmology. Scattered throughout his 41 sermons of Lent 1509 are descriptions, anecdotes, explanations, and comments which assemble to articulate a comprehensive representation of several elements of the natural environment, including animals, plants, general reasons for inclement weather, and suggestions for addressing the latter.

To begin: along with the metaphor of the ostrich egg mentioned above, during his third sermon of the cycle, Geiler preached about the egalitarian nature of an ant colony. Whether personally lacking knowledge about the presence of a queen in every colony or choosing to follow Nider in this, Geiler portrayed ants as having no leader except God Himself. With God's leadership accepted by both solitary creatures, such as lions or bears, and social creatures, such as ants, swallows, sheep, herring, or bees, Geiler preached about the need and motivation for human beings – included in the category of social creatures – to accept God's leadership. As part of his explanation for the ants' obedience to God, Geiler gave his understanding of God's role in the creation of animals.[51]

The preacher began by reminding his audience that God is the common leader of all creation, and excepting for humans, who exercise free will, divine rule may be seen everywhere on earth. That is, no worm may wriggle or crawl without God's help and direction, as, when a creature is made, God gives to that creature an impulse of energy, by which energy they must live their lives. To further explain this understanding of the manner in which creatures gained life, Geiler then offered an interesting metaphor, whose source domain was archery. That is, says the preacher, when someone wishes to shoot at a target with a crossbow, he lays an arrow at the breastplate and pulls the string tight; as the string hits the end of the bolt, then the bolt is driven towards its target by the same impulse. In this same manner, the preacher told his audience, God bestows life-energy upon a creature when he makes it and by the force of this energy, the creature generates its behaviour. The eagle flies in this way, the hare leaps, and the horse runs. Thus, Geiler asserted, God also governs all the ants. He further clarified that he was not discussing the common governance of all things, but the leading of their souls.[52]

God's actions were made very clear to listening audiences: God creates things and provides the means for his creations to move about according to the manner in which they were created. This passage is not found in the corresponding chapter of Nider's *Formicarius*, which begins and concludes by lauding the lack of visible leadership in ant communities, but where, for the Dominican monk, the lesson from the ant was confined to the proper management of monasteries and cloisters.[53]

This fascinating passage can be queried in a number of ways, beginning with an exploration of Geiler's understanding of the process through which God makes animals. First, God creates something; secondly, He gives it life. The Vulgate account of humanity's creation, in Genesis 2:7, speaks to God's formation of man using material described with the word *limo*, which can be translated into English as *mud*, *clay*, or even *slime*, while the same term in Greek is translated as *dust*. Geiler would have been familiar with the Vulgate, but Erasmus had not yet published his Greek New Testament in 1509, and Geiler, as mentioned earlier, did not learn Greek and so did not browse through the Pentateuch in that language. Geiler's words, consistent with the Vulgate Genesis and following immediately after his reference to wriggling, crawling (cold and slimy) worms, evoke an image of a potter or sculptor crafting animals from a wet, sticky substance and then energizing them to a leaping, running, flying life. Echoing the Biblical sequence of events, the hierarchy of creation is maintained: before God is able to introduce life to His creation, He must create a material receptacle.

The impression of God as a potter or craftsman, however, is quickly superseded by Geiler's surprising depiction of God as an archer. While Adam is gently brought to life through God's breath, the Creator provides life energy to the eagle, hare, and horse (as well as, presumably, all the other non-human beings) in the manner that an archer sends an arrow to the target. The construction of any metaphor combines individual inspiration with culturally accepted associations and similarities, and the portrayal of God as an archer is uncommon. In Geiler's attempt to communicate the manner in which creatures are given mobility and life, he may have relied upon his audience's ready knowledge of the weapon. Strasbourgeois of the early sixteenth century were acquainted with firearms; this generation saw their introduction and the city owned a canon named *der Strauss* (the Ostrich). However, while artillery was manufactured throughout Alsace – that of Strasbourg was particularly renowned – firearms were expensive and secrecy surrounded their manufacture. Restricted access to gunpowder limited its customary use to the military and related guilds, making it more likely that those outside these professions would use and/or understand a crossbow rather than a firearm.[54]

This metaphorical placement of God in relation to His creatures, as creating them and then sending life energy to them, defines Him as a transcendental divinity, which is to say, outside His creation. God is not present in the animals, whether eagle or ant; instead, He creates them, He vivifies them, and since they accept His leadership, He governs them. Each creature lives according to His design.

Three days later, on Invocavit (the first Sunday of Lent), Geiler launched into the sixth sermon by identifying the source of the wisdom of animals, giving a substantial expansion upon his earlier sermon on the gift of life. The particular quality of ants upon which he expanded was their wisdom, which he developed into four categories: awareness, wisdom, reason, and astuteness. Ants are untiring, he preached, and they have a good recollection of what they are supposed to do; this leads them to shuffling back and forth, from early in the morning until late, and they accomplish their work. Creating an imaginary dialogue between himself and an audience member, Geiler pretends the listener challenges him by asking how ants can possess reason, as the imaginary questioner is accustomed to reason only in humans. Is Geiler suggesting, he speculates about himself, that animals also have reason? It is true, he responds to the imaginary interlocutor, that if actually talking about reason, then men alone have reason. But, he goes on, he is using the word 'reason' in the sense of its common understanding: astuteness, and the careful attentiveness an animal possesses. He advises his real listeners to look for the wisdom of God in the business of the animals, asserting that astuteness was placed into both a tiny gnat and a large elephant. Returning to the ants, Geiler draws attention to their eyestalks, saying that within these little spikes God has placed His astuteness, which the preacher identifies as memory and reason.[55]

Geiler goes on in his sermon to extend the example from ants to the other animals such as sheep, whose young are able to show their capacity to distinguish between an ass and a wolf (when never seen before) by fleeing the wolf but not the ass. Since enmity is invisible and inaccessible to the senses, he asks, who tells the lamb that the wolf is his enemy and not the ass? He finds the answer in the knowledge and astuteness given to the nature of the animals by God, which he establishes as their ability to reason.

Geiler then describes the manner in which the wisdom of animals draws near to the wisdom of men, as much as they are able, as that of men draws near to the wisdom of angels. Creatures, he says, are like a chain, where one ring grips the other ring, the second ring grips the third, and so forth, and where in this way, all creatures are linked together. Even trees go towards the reason of man as much as they are able. Geiler provides his proof for this concept in the behaviour of creatures: the success of spiders in hanging threads and their placement, the manner in which mother cats hide their kittens from tomcats, or how swallows make their nests. He considers their behaviour to be a result of their God-given reason, and that, in this context, the ant is smarter and has a better memory than many other animals.

Whereas in the sermon of three days earlier, Geiler simply presents life-energy as the common gift God directs into each animal, which then behaves according to its design, this passage expands and complicates the relationship between God and creatures by adding the quality of wisdom to the energy of life. Both are placed in the creature by God, and, in the ant's case, rather delicately at the top of its antennae. It is explicitly God's wisdom which has been inserted into each creature; that is, the wisdom belongs to God, and is not inherent in the creature nor its design. With this passage, Geiler shifts authority over the behaviour of animals from each individual creature (acting in accordance with its design) to God, whose wisdom it is which motivates them to do what they must to survive, like running away from wolves. While God remains conceptually transcendent to His creatures, a particular quality of His (wisdom) now dwells in each eagle-, hare-, or horse-shaped material receptacle, along with the energy of movement He provides.

Moreover, Geiler presents the shared quality of God's wisdom as the common element which unites God's creation, including trees, animals, human beings, and angels, each one approaching or 'reaching to' its wiser neighbour. What follows is a sketch of the Great Chain of Being, an influential concept much debated in feudal Europe. The organization of all creatures as linked together in a vast hierarchy was a concept initially developed by Plato, Aristotle, and Proclus; the Christianization of the *scala naturae* took place following its translation during the twelfth century, and was substantially expanded by Thomas Aquinas in several of his works. Central to the concept is the view of God as principal cause and measure of all things, and the more an Intelligence (a creature) approaches God (*accedit*), the more he/she/it participates in God's goodness and life. The organizing concept is that God is good in substance and exists through Himself, while creatures are good and exist by degrees of participation in God. That is, creatures were understood to take their place in the hierarchy of being through proximity, as they approach or recede from Him. Controversy raged over the distance of God from creation, with opponents arguing about how God's nature influences His approachability. By alluding to the Great Chain of Being, Geiler is describing the relationship of creatures to one another and to God, not suggesting the physical mobility of trees.[56]

Equally interesting to Geiler's understanding of the creation and behaviour of animals are his reasons for unexpected or unhappy human interactions with the environment, such as wild animal attacks. For these, he offers a nuanced analysis which combines ordinary bestial motives with diabolic and divine will. On Oculi, the third Sunday of Lent, his twentieth sermon included an exposition of the seven distinct reasons which might motivate a wolf attack on a human. The first reason is circumstantial, such as when winter is upon the land and other prey cannot be found; the second is when the wolf is *grimme* (wrathful or furious), such as when its young are threatened. The third, fourth, and fifth are due to old age and debility, the easily tempting availability of a human, and from its own foolishness, similar to a dog chasing its own tail. The sixth type of wolf attack occurs when the devil assumes the form of a wolf; the seventh destructive wolf is God's ordaining, where God chooses to punish some land or village through a wolf. Not every wolf attack is diabolically motivated, nor every bad dream, as Geiler explains during his fifteenth and sixteenth sermons detailing the reasons for various types of dreams. While he does not turn his analysis to extreme weather events during the 1509 Lenten sermon cycle, given the context, it is reasonable to extrapolate from wolf attacks and dreams to the weather and assume that, in Geiler's view, not every thunderstorm was caused by the devil.[57]

Notwithstanding mundane reasons for extreme weather events, Geiler accepts that magic does have a role to play in the creation of storms or hail. However, in *Die Emeis*, the preacher's views of magic do not feature spells or curses directly cast by witches and wizards in a straightforward manner, *à la* Harry Potter, where the ability of the practitioner directly causes the desired effect. If this were the case, the ability to use magic to break or bend the laws of nature created by God would position human spell-casters as equal to God in their authority over the natural world. This idea treads upon very dangerous theological terrain, shares obvious similarities with Lucifer's hubris, and Geiler is steadfast in correcting it. Rather, he repeatedly presents a complex dynamic wherein a human makes certain signals or signs to the devil indicating malicious intent; the devil's own God-given powers allow *him* to either do the work himself or to dispatch a demon to fulfil the human's desire.

For example, his twenty-third sermon, on the Tuesday following Oculi, focuses on an exposition of the methods through which the devil is able to inflict harm upon humanity and Geiler includes a brief explanation of how he understands magic to operate with respect to destructive weather. When witches or sorcerers are seen sweetly brushing a manikin or hitting a bundle of straw on their leg, he says, this is not the principal cause of hail and thunderstorms, but only the prompt which brings about the actual deed. Such things, Geiler articulates clearly, are nothing more than a signal summoning the devil, who arrives and undertakes to fulfil the desire of the human.[58]

Noting again the spiritual hierarchy, Geiler explicitly diminishes the role of both the witch and the straw doll in creating hail and thunderstorms, instead making the role of the former into that of a summoning agent. It is the devil himself who actively creates the hail and thunderstorms. The preacher repeats his point the next day, emphasizing the devil's desire for equality with God as the foundation for this dark parallel sacrament. This desire is Satan's traditional motivation, of course, for rebellion against God. Geiler, however, is clear in distinguishing between God's sacraments, which are reliable when used as intended, and the devil's version, which are not. A witch throwing water over her head with a broom, then, is not causing rain; she is activating her pact with the devil, whose divinely ordained but wilfully misused power is the true cause of the storm. Since the devil, however, is not as powerful as God, witches cannot be assured of success. Strikingly, Lucifer's unreliability is Geiler's most significant caution against consorting with the devil throughout *Die Emeis*. The human who would initiate diabolic action is described frequently as malicious, corrupt, spiteful, or evil, but no mention is made of eternal damnation or purgatory's purifying fires as terrifying consequences for summoning the unholy. It is possible that the devil or his demons may arrive and initiate misfortune without human solicitation (perhaps an aspect of their unreliability), as mentioned in his eighteenth sermon, but Geiler presents the summoning of demons and their master as a relatively easy action to take, perhaps due to the swarming clouds of evil spirits at work in the world (like mosquitoes).[59]

As well as swarming about in annoying clouds, Geiler portrays the demonic world as hierarchically ordered; evil spirits are ranked in terms of power and ability, just like humans. Some, such as these mentioned during Geiler's twenty-second sermon, are among the weaker ranks; Geiler relays that some demonic spirits who did not receive great power from God at the time of their creation are only able to affect small annoyances, like rattling bowls or cluttering a house with trash. While smaller spirits may rattle the bowls, though, the devil himself is able to unsettle and terrorize an entire village. Geiler shares an anecdote about the calamities of a village in the diocese of Mainz, where the devil made an entire village unsettled by hammering on houses, breaking windows, and bringing about many mishaps and general discord. Narrowing in on an individual mute man (a good soul), the devil burned down his house; when the man moved to another house, the devil burned that one down as well. Since he could no longer find shelter in his village, the mute man was obliged to live in nearby fields. Priests organized a rogation, but during it, the devil cast a stone at someone and wounded their head. As his final *tour de force*, the devil destroyed the entire village by burning down all the houses.[60]

The level of threat posed to humanity by the unholy varies according to the strength which God had given to any particular angel at the moment of its creation. Geiler does not address the manner in which angels were brought to life in the sermons of *Die Emeis*, nor does he discuss their fall from grace therein. Irrespective of the hierarchy in demonic society, however, there are clear limits to their ability; specifically, Geiler represents the devil and his demons as unable to transcend nature's limits as established by God. This includes the transformation of a human into an animal, or one animal into another; since this would not occur within nature, the devil cannot bring it about.[61]

Unable to create *ex nihilo*, then, Geiler presents the devil as rather helplessly left in the position of requiring God's materials to implement diabolical design, a sort of supernatural domestic terrorist. The preacher emphasizes this point again during his twenty-first and thirty-third sermons. During the former, he says that the devil is able to make the animals which can be made by nature or which are generated by putrefaction. He draws the parallel with mice, snakes, or frogs, which grow in the rubbish people leave lying around, and posits that if the devil has the seed of snakes, of frogs, or of other incomplete animals, he can breed them there with the warmth that is generated. During the latter, Geiler asserts that the devil is able to make snow, rain, wind, hail, and thunder because he can assemble the four humours (earth, air, water, and fire) for a brief period – but only after being given the signal from a witch. This diabolic capacity, Geiler asserts, means that witches are able to generate a hailstorm in a heated room – but only if there is water present, as diabolic creation must have source materials with which to work.[62]

Geiler does not leave his Lenten audiences without recourse against these bowl-rattling and storm-bringing demons. In his thirty-second sermon of Lent, delivered on the Friday after Laetare, the fourth Sunday of Lent, he suggests three potentially useful instruments against the devil's manipulation of atmospheric conditions: weather blessings, the ringing of bells, and legislation. The first directly calls upon God to exorcise bad spirits, the second takes advantage of sacred sound, and the third relies upon legislation to punish and execute witches. He also mentions an approach from Gaul and Swabia which was also practised in Alsace: speaking selections from the Gospel of John to the storm. Geiler provides the source of authority for each (Book of Job; Old Testament; the Emperor, and the Gospel of John, respectively) and emphasizes that calling upon God is the strongest remedy available to humans for magically created bad weather.[63]

These were familiar tools to Geiler's parishioners and they would have been well-known in the community as means of overcoming the extreme weather events manifested by devil and his demons. As mentioned in the previous section, bells were a common element of public space during the medieval period. Widespread and contested, the Roman Catholic practice of exorcism could be found at all levels of religious pastoral care, from the municipal to the individual. The practice was so prevalent, however, that many rituals to cast out evil spirits were unauthorized and improvised, leading to the suspicion that they were ineffective or, worse, served to summon the very spirits they were intended to remove. However, the invocation of Saint John by speaking aloud the first words of his Gospel, as recommended by Geiler to his audience, was one of the more common practices of exorcists and, according to Sarah Ferber, an element of church magic much favoured since at least the turn of the millennium.[64]

Based on sixth-century Justinian law codes, legislation against witchcraft during Geiler's lifetime had been formulated a millennium before, then rediscovered and reintroduced to Europeans during the twelfth and thirteenth centuries. It was supported and guided by a series of Papal decretals and bulls which increasingly emphasized the heretical nature of using magic. These culminated in 1326 with the blanket excommunication of anyone convicted of involvement with demonic invocation and demon worship. By the last half of the fifteenth century, these religious views of magic and witchcraft – which had gained prominence with William of Auvergne, bishop of Paris from 1228 to 1249, and Thomas Aquinas – found social authority and coalesced into laws which identified such behaviour as a crime and provided for its punishment. Uniting the authority of the Latin Church with civil legislation against unorthodox spiritual activity of any kind, a combination of the inquisitorial procedure and the introduction of judicial torture in the 1400s had created a legal process whereby an inquisitorial tribunal was responsible for investigating accusations of witchcraft and for persecuting the case. Magistrates could also initiate investigations based merely on reports or rumour, without any formal accusation having been laid.[65]

Witchcraft trials of individuals or small groups were held in at least six towns of western Switzerland and southwestern Germany during the last decades of the fifteenth century. Works such as the *Malleus Maleficarum* urged for more widespread and intense persecutions, but trials for witchcraft declined in number even in those areas where they had sprung up most quickly, eventually petering out by 1507. Legislation, however, like that of Maximilian I mentioned by Geiler or of the Bishop of Bamberg, whose new laws issued in 1507 justified the use of torture and execution for the crime of using magic, remained active. These laws were frequently invoked during the second and greater wave of accusations and executions in the latter half of the sixteenth century, during which, according to research by Wolfgang Behringer, there was a correlation between extreme weather events and scapegoating activities such as witchcraft trials and witch burnings. With his appreciative inclusion of legislation against witchcraft as a suggested remedy for storms, Geiler is expressing his approval for the punishment of theological crimes by civil authorities, despite the separation between clerical and lay communities he generally preferred.[66]

This case study has described representations of nature made by Johannes Geiler von Kaysersberg during his 1509 sermon cycle published as *Die Emeis* and explored their implications. Despite the absence of any mention of current weather events or their consequences for his fellow Alsatians, by Good Friday 1509 he had managed to remind regular audience members that God created each being and placed His wisdom therein, that not all untoward encounters with the non-human world were diabolical in origin, that human witches could not directly cause bad weather and who was, in fact, responsible for some of it (the devil), and how to address extreme weather events when they occurred, among other things.

In reminding his audience of these orthodox Catholic tenets about nature, Geiler's primary frame of reference centred on the *scala naturae*, a concept which understands the natural world as hierarchically ordered according to each being's proximity to God. Geiler did not create this frame of reference or representation of the natural world himself, but learned much of it during his education and professional development. The next section explores the intellectual and theological context from which the orthodox preacher approached his topic.

Natural philosophy, the supernatural, humanism, and Geiler

Geiler's prominence as a reform-oriented Catholic preacher shortly before the onset of the Protestant Reformation has sharply focused scholarly attention on his theological relationship to the upcoming religious turmoil, as mentioned earlier. However, just as Rita Voltmer drew attention to hitherto unconsidered political aspects of his career in the pulpit, the role of natural philosophy in his education and his preaching has gone unexamined until now. Developing a deeper understanding of the manner in which he represented the natural environment is reliant on knowledge of the intellectual and theological traditions and debates in which he was educated and from which he preached. This section of the chapter describes the development of some ideas about the natural environment that were dominant in the early sixteenth century and explores Geiler's position with respect to them.

As summarized by Edward Grant in his volume *The Nature of Natural Philosophy in the Late Middle Ages*, an orthodox university-trained Catholic theologian of the late fifteenth century would have enjoyed, perhaps surprisingly, a thorough knowledge of Aristotelian natural philosophy. The inception of translations of Aristotle's *libri naturales*, *Metaphysics*, and other works from Arabic and Greek to Latin in the mid-twelfth century had magnetized the attention of European scholars, and the dramatic (re)appearance of this rich vein of Classical philosophy coincided with – and supported – the establishment of the university as an institution. Once accepted for admission, undergraduate students all shared a common course of study in the Arts Faculty before specializing, if accepted, in one of the higher degrees of theology, law, or medicine. The medieval Arts curriculum quickly came to lean heavily on Aristotle's works about natural philosophy and logic, and students trained exhaustively to master a six-step format interrogating natural phenomena from the perspective of the Classical Greek philosopher.[67]

The unlimited and imaginative range of early theses which were produced during the twelfth century included those whose conclusions conflicted with Church teachings. In particular, the Aristotelian method sometimes brought exponents to logical opposition with orthodox doctrine about, for example, God's capacity to overcome the materially impossible, the existence of other worlds (such as heaven), or the eternal nature of God. These unexpected and unwanted intellectual confrontations repeatedly provoked religious authorities to attempt bans on the reading, discussing, or teaching of Aristotle, which culminated in the Condemnations of 1277 issued by Etienne Tempier, Bishop of Paris. Nevertheless, however explicitly the prohibitions articulated the threat of excommunication, the measures failed to prevent the penetration of Aristotelian natural philosophy and the philosopher's analytic methods into Arts faculties, including at the influential University of Paris.[68]

During this campaign by the Roman Catholic Church for intellectual control of the educated elite in the high middle ages, another overt and more effective attempt to dam the application of Aristotelian methodology to matters of faith was the oath required of Arts masters that they avoid the introduction of theology into their interrogations, initially required at the University of Paris in 1272 and soon elsewhere. Questions about God's omnipotence, His essential nature, or the truthfulness and accuracy of Catholic doctrine were to be avoided, or, if perchance theology or questions about God did appear, the oath demanded that any issues were to be resolved in favour of

God's supremacy as articulated by Catholic orthodoxy. Perhaps due to an intellectual predisposition for avoiding supernatural explanations for natural phenomena, masters were happy to exclude theological questions and issues from their Arts curriculum. However, since medieval Catholic doctrine included clear representations of God's material Creation, a consequence of excluding God from the construction of theses was that direct observation of the natural world as a way of proving or disproving a hypothesis was also excluded. Rather than suppressing natural philosophy, the oath served to create a situation where generations of university graduates concluded their Bachelors and Masters of Arts with a solid foundation in Aristotle's quantitative and analytical methods as only *speculatively* applied to natural philosophy. If the student should then progress to the doctoral stage, those methods would be brought to bear on theological, medical, or legal questions. The body of scholarship produced in this manner became known as Scholasticism.

For the post-graduate medieval theologian interested in exploring questions about God's Creation, two means of conceptualizing His power gained prominence in mid-thirteenth-century Paris and provided a way of measuring it while simultaneously guaranteeing His freedom: the *potentia absoluta* of God (the unlimited power of God to manifest His will) and His *potentia ordinata* (the subset of choices made by God in creating nature). Essentially, these concepts provided a way of comparing hypothetical reality and actual reality; what were the choices God had made to create the earth, for example? Since God ordained that reality, both hypothetical and actual, would function within the bounds of logic, logic itself became the primary method of analyzing and understanding God's will.

The prevalent analytical approach became known as *questiones*, where first a theme was introduced and then a question established, such as whether there are, or could be, more worlds than ours or whether the universe exists in eternity. If the question's resolution led to a logical contradiction, God could not perform the hypothesis; if no contradiction was involved, there would be no impediment to this particular manifestation of God's power. In *questiones'* oral form, the master posed and answered the question himself, with any objections spontaneously raised by auditors to be addressed on the spot. The student was expected to participate either as a respondent to objections raised or to determine and resolve a question himself. For university masters of any subject, this format was a regular feature of intellectual life. Although the imaginative flourishes of hypotheses questioned by medieval scholars became a source of mockery to intellectuals of the Enlightenment, this application of logic to intellectual matters was the primary feature of the scholastic method and remains a primary feature of intellectual activity today. Moreover, an unprecedented level of imaginative interrogation into potential or alternative realities for something observed in the natural world resulted from this approach, and such an expansive flight into imagination may have been the foundation of empirical inquiry when conducting it became culturally acceptable in later centuries.[69]

Two significant drawbacks to the scholastic method were an inability to resolve an argument, and the atomization of Aristotle's treatises. Regarding the first, scholars relied on established authors for arguing their theses; settling a dispute relied on consensus, as rare in late medieval Europe as today. Extracting unique statements from Aristotle's works for elaboration and questioning, however, may have exerted a more depressing influence, as doing so inhibited the syntheses of issues into conclusions and stymied investigation into the Aristotelian worldview for inconsistency, deficiencies, or weakness. For assistance in constructing their theses, scholars called upon respected authorities from the Church or elsewhere, which led both to the unquestioning perpetuation of error and endlessly repeated circular arguments.[70]

Largely as a result of the Scholastic method, late medieval theological discourse developed towards abstraction. Hypotheses, conclusions, and refuted arguments were an important part of demonstrating solutions provided by a God-given reason alone within the framework of faith; they were thought experiments designed to refute or uphold a theory. The question of the relationship between a concept and the material example of that concept became hotly debated, likely due to the contribution its resolution would make towards validating increasingly abstract and obscure speculations. The issue came to be sharply focused on the question of universals, which is to say, on the question of whether or not universal concepts like 'green' possessed an existence independent of the human mind and, if so, how this related to their material subjects, like trees or grass. One perspective was that shared by scholars such as Peter Abelard or Roscelin, who referred to universals as 'concepts created by the mind without extramental referents.' That is to say, as being created purely within the mind, such concepts could not, therefore, possess any inherent relationship to external reality.[71]

This view, initially referred to as *terminist logic*, eventually became associated with the term *nominalism*. In his detailed study *The Intellectual Origins of the European Reformation*, though, Alistair McGrath notes that the term seems to have disappeared from use during the thirteenth century, rendering its application to scholars of the 1300s such as William of Ockham or Gregory of Rimini an anachronism. The question addressed by the term did not disappear, however, nor did a scholarly predisposition to refute the reality of universal concepts. Instead, the extrinsic relationship of concept to reality became part of a larger movement in late medieval theology known as the *via moderna*, which, as characterized by McGrath, was based on broad and loosely defined common epistemological assumptions shared by prominent individual scholars and those under their influence. Most particularly, *moderni* shared a confidence in the methodological approach of logical criticism for understanding God and his Creation, taking advantage of the speculative opportunities provided by the distinction between the *potentia absoluta* and the *potentia ordinata*, along with the recognition of God's freedom in behaving reliably when exercising his will regarding Creation. This last view was known as *voluntarism* (in that God voluntarily chose to uphold his agreement with human beings), and led to the merciful view that doing one's best, or *facere quod in se est*, was rewarded with the gift of justifying grace.

Voluntarism's theological implications were particularly relevant when it came to religiously meritorious action, where an awareness of the distinction between a moral act, committed by a human being, and God's acknowledge- ment of that action as religiously meritorious led to doubt about the eternal efficacy of the sacraments or moral behaviour for winning human salvation. Scholastics of the *via moderna* come to consider God as having voluntarily and charitably chosen to be reliable in awarding merit to specific deeds. This led to a view where receiving a sacrament inspired an act of charity, or grace, from God, as he voluntarily fulfilled his promised will in bestowing the bless- ing. With respect to natural philosophy, voluntarism meant that the natural world could be relied upon to continue as God had created it, notwithstand- ing God's freedom to miraculously intervene in normal operations.[72]

In contrast with the view held by *moderni*, scholars of the *via antiqua* con- tinued to agree with Aquinas' expansion of Augustine's concept of a created habit of grace which, as a supernatural element in the soul, was understood to mediate the human relationship with God. In this context, acceptance of the sacraments was the crucial means of bringing the human soul to its maker by changing the very constitution of the people who received them into something which was able to approach God. That is, an external force (the sacraments) was applied to fallen human nature which allowed it to be acceptable to God; God's purity prevented Him from approaching humanity on His own. Other components of the *via antiqua* included an acceptance of linguistic structures as part of the essential nature of things, independent of human convention or grammar. This view, known as modist logic, modistic logic, or speculative grammar, attempted to understand language and reality through three 'modes': *modi essendi* (the way of being), *modi intelligendi* (the way of understanding), and *modi significandi* (the way of meaning).[73]

Philosophical methods and perspectives as varied as the *via moderna* and the *via antiqua* were entirely acceptable within the late medieval Catholic Church. Intellectual freedom offered to theologians was not yet hampered by political divisions supported by military action, as would occur in the sixteenth century and afterwards. While disputes between terminalism and realism or about the nature of justification might roil, proponents were free to determine their opinion in each dispute with a relative independence from political considerations, as all acknowledged the primacy of the Pope and Roman Catholic Church. Universities might become known for teachings associated with a particular *via*, such as Cologne's association with the *via antiqua* and Erfurt's with the *via moderna*; Geiler's *alma mater* at Basel was one of many where theologians from both *viae* associated comfortably.[74]

The relationship between the theses of natural philosophy debated at me- dieval universities and the theoretical knowledge of nature among the laity was primarily mediated by preachers; Geiler's attempts to educate his secu- lar audience with insights from his theological training were not unusual.

Alongside the administrative and political training whereby universities served the agenda of their founders, mediation of learned knowledge into surrounding populations became an important consequence of their establishment. As declared by Hans-Jocher Schiewer, 'In the late fourteenth and in the fifteenth centuries universities became more and more service institutions which – through their academic teachers – authorized, mediated and distributed theological knowledge and sermons in the vernacular.'[75]

Johannes Geiler was no exception to this profile of the late medieval theologian learned in Aristotelian natural philosophy. We know from Jakob Wimpfeling, for example, that the two met when Geiler was instructing classes on Aristostle's *oeuvre* as part of his university career (Geiler being five years senior to Wimpfeling). Geiler's sermons in 1509 regularly and casually refer directly to the philosopher; knowledge of Aristotelian natural philosophy is found throughout *Die Emeis*.[76]

Even the basic format of Geiler's sermons is structured similarly to the *questiones* method used by university theologians trained in natural philosophy: following an introduction of the theme, Geiler poses a question, and then answers it with lists of supporting evidence. The questions he asks are variations on the theme of personal improvement – 'What can we learn from this?,' or 'What should I learn from this?' – a form suitable for preaching, rather than debating. He regularly ascribes the interjection of questions or objections to his audience or responds to actual unrecorded comments, which give him the opportunity to clarify his previous point.[77]

It is challenging to situate Geiler's stance with respect to the two significant theological approaches to natural philosophy of the late medieval period, since recent scholarship positions individual scholars according to the methodology taken in their written publications. Geiler, however, was self-consciously transmitting the knowledge of theologians as a preacher, rather than developing it further as a natural philosopher, which he would only have expected to do as the occupant of a position at a university. The change in his career path from academia to the pulpit drew his focus away from active scholarship, and because his surviving works do not show him disputing or arguing, this makes his personal method of approaching questions and issues obscure. His position *vis-á-vis* contemporary theological schools of thought, then, must be discerned from whose arguments he chose to preach and whose he ignored.

Both E. Jane Dempsey Douglass, author of an in-depth exploration of Geiler's theological views in her volume *Justification in Late Medieval Preaching: A Study of John Geiler of Keisersberg*, and Thomas A. Brady, Jr., understand Geiler to have been a traveller on the *via moderna*. Douglass points out that Geiler's preaching on the nature of sin and free will serves to illustrate this conclusion, as do his views on the role of grace in human salvation. Douglass' survey of Geiler's surviving works concludes that the preacher mostly agrees with Biel that *facere quod in se est* is sufficient to receive God's grace, and that free will allows for this to happen if chosen. In *Die Emeis*, Geiler's perspective on the question can be seen in his faith that God will reliably bestow the blessing sought by any particular sacrament, his confidence that the natural world will continue as God designed it, and the sharp contrast he makes with the Devil's unreliability.[78]

As well, Douglass finds Geiler's broad-minded spirit of inquiry, prepared as he was to transcend the schisms and party struggles of late medieval theology, to be characteristic of the *moderni* and in contrast to the suspicious approach of the *antiqui*. Geiler maintained this spirit even at the end of his life, when, at 63 years of age in 1509, his ninth and tenth Lenten sermons focused on the communal nature of ant society and the absence of impediments erected by ants for one another in achieving their goals. Particularly in the tenth sermon, the essential unity of the Christian church was explicitly emphasized with comments like 'whoever has something good to say is for me a good teacher,' accompanied by sharp criticism of unseemly and useless bickering between schools of thought.[79]

Johannes Geiler von Kaysersberg, then, can be considered a *moderno* in the context of late medieval theological schools; his professional reputation as a learned scholar rested upon the depth and breadth of his knowledge in that Aristotelian-shaped field. However, Aristotle's *Meteorologica* (translated into Latin by Gerard of Cremona in the twelfth century) did not provide an adequate explanatory framework for understanding extreme weather events. In it, the Greek philosopher explains the operations of rain, tornadoes, thunder, and lightning, but does not include a rationale for their occasional expansions into extreme weather events nor a meaningful understanding of any ensuing disaster. To explain the causes of some natural disasters and to provide a way of addressing them, Geiler turned to spiritual agents in his sermons: Satan and his demons, generally, but also God and His angels. In a move that highlights again the many differences between natural philosophy of the early sixteenth-century and twenty-first-century natural science, Geiler preached from the cutting edge of late medieval natural philosophy not only by including the role and influence of the supernatural on nature in his sermons, but by carefully working out the logical implications of demonic and divine activity in a natural world created by God, operating with His wisdom and according to His design. While explicitly valuing the use of reason in understanding revelation, thereby illustrating once more his orientation towards the *via moderna*, Geiler and his most respected contemporaries 'thought with demons.'[80]

Demonic participation in earthly life, along with the participation of God and His host of angels, was both taken for granted and unquestioned by late medieval Catholic Europeans. Late medieval theology comprehended Lucifer's fall as a punishment which deprived him and his associates of God's grace, but which jailed them on Earth and left their preternatural essences largely intact. The devil and his demons were understood to be impotent before the laws of nature, as only its Creator had the ability to transcend the order He had created. They did have, however, ancient and complete knowledge of the natural world, having been present at its creation, as well as great strength and fantastic speed. These relics of their angelic nature made demons superior to humanity in terms of their knowledge of the environment and their ability to manipulate it. Objects, for example, could appear to have changed location instantly through the intervention of a demon, bodily humours could be disturbed by the same, leading to illness, or the weather could be directly affected. Two other factors complicated the subject: unlimited powers of deception were ascribed to the devil, and the natural world itself was capable of both producing wonders and appearing to produce wonders. For any event, then, theologians were faced with the need to discern between real demonic effects, illusory demonic effects, real natural wonders, and illusory natural wonders, or any combination of the four.

It is important to underline that while demonic and diabolically influenced events were considered distinct from the order of nature as created by God, they were still categorically understood to be incorporated into the natural world: the devil was constrained to operate within the sub-lunary sphere as part of his punishment. He could create *mira* (wonders), but *miracula* (miracles) could only occur through the will of God. In the late medieval distinction of natural and supernatural, only God could be understood as supernatural, due to His position as the creator of everything else. All of creation, that which was created, is, therefore, natural, and this included the devil and his fellow fallen angels, as well as angels who remained loyal to God in heaven. Thomas Aquinas exerted a great influence in the establishment of this view in the thirteenth century, when he asserted that God alone could perform miracles. Aquinas went on to say that although it may seem that a creature is performing miracles, they are not true miracles but performed through hidden natural forces, like the pseudo-miracles of demons.[81]

Distinguishing a *mira* from a *miracula* required reliable knowledge about natural causation, research into which was conveniently conducted in the late medieval period by scholastic theologians trained in Aristotelian natural philosophy, as described above. Several of Geiler's sermons during the 1509 Lenten cycle show the precision of his demarcation between the natural order of events and an unholy interference in that order. His comments about the devil's need for the seed (semen) of an animal in order to create more of them illustrate this precision, as does his assertion that the devil can only make a storm if water, the basic material, is already available. The devil and his demons may be able to manipulate material objects and the natural world to threaten people and destroy property, but Geiler explicitly warns that the devil cannot make something come to pass which is not already part of the natural order.

When communicating conventional wisdom about the non-human environment, then, Geiler drew on his training in both natural philosophy and theology to develop his argument with respect to diabolic action in a divinely created environment. As noted by Clarke, the demands of late medieval Christianity were that the devil be strong in relation to humanity and weak in the face of God's power. Two brief examples: first, as described earlier, Geiler's rejection of wolf attacks as universally motivated by diabolic agents. Second, his careful sorting of the actors at work in the creation of a destructive hail-storm: the witch may be malicious and have traffic with the devil, but humans do *not* have the ability to exercise supernatural power. If threatened by diabolical events, human beings are advised to restore the natural order by calling upon their stronger ally, God, to exorcise and dispel the unholy spirit at work.

In the sermons compiled as *Die Emeis*, when he recommended ringing bells, exorcising the clouds, and legislating against the summoning of evil spirits, Geiler was educating the listeners in Strasbourg's *Liebfrauenmunster* about widely held and theologically sound methods for securing good weather. In the context where Geiler was reasoning from a sixteenth-century faith-based standpoint, and not from the perspective of twenty-first-century scientific materialism, his comments about the natural world are perfectly rational, and he repeatedly confronts views which, in his opinion, are irrational because they are not in concordance with natural philosophy or Scripture.[82]

Geiler's exploration of witchcraft and magic in *Die Emeis* is not limited to its use with respect to the natural world. As well as during Lent of 1509, Geiler made a number of assertions on the topic during his 40-year career, listed by Rita Voltmer: that the devil was very much available, even eager, to compact with humans; that sexual relations between humans and demons were possible, as was the impregnation of women by demons with stolen human semen; that the notorious night flight, whether by women to a gathering of witches or by men to the mountain of Venus, whether willingly or unwillingly undertaken, was both a dream vision and bodily possible, but in any case, depended upon the devil's powers for execution; that women participating in a night flight with pagan goddesses or fairy folk was a diabolic illusion performed upon the borrowed bodies of sleeping women; that the devil attempted to harm human bodies and souls through diverse strategies, including the introduction of illness to humans and animals, as well as through bad weather, and that all such damage was only undertaken with the permission of God; that any damage inflicted through sorcery should on no account be annulled though the use of further magic, divination, or magical conjuring, and that only authorized remedies, such as prayer, crossing oneself, confession, or sacramental objects were to be used; that the success of exorcism depended upon the purity and piety of the exorcist; that human metamorphosis into animals was impossible; that most of the phenomena attributed to magic (displacement by spirits, visions, changelings, witches riding horses to exhaustion, love charms, or knocking spirits) were illusions or spells cast by the devil, or simple frauds; that the vulnerability of women to diabolic sorcery was due to their inherent nature, but that this nature also gave them a greater understanding of creation, so that women were both the most pious and the most evil of humans; and, finally, that the reason God gave such destructive power to the use of the devil, sorcerers, and magicians was to drive the wicked further into diabolic perversion so as to test the pious and the good – although this did not excuse civil authorities from prosecuting heretics, blasphemers, and magicians as appropriate.[83]

Alongside intellectual trends and assumptions such as these in natural philosophy and theology, late medieval Europe also saw the development of Humanism as a philosophical stance. Northern European humanists found common ground in three ideals, methods, and presuppositions: a literary or cultural programme focused on the ideal of good literature, a religious programme directed towards a renewed Christianity, and a political programme working towards peace in Europe. Geiler's vocation brought him into distinct alignment with the second and it is unlikely that he disagreed with the other two. His goal of a refreshed Christianity, though, can be understood as related more to his sense of responsibility for the Church in general, and for his community in particular, than any donning of a humanist identity, based as such an identity was on a self-conscious participation in a linguistic and literary network with which Geiler did not engage. Indeed, Thomas A. Brady, Jr, declares that Geiler is more attuned to Blickle's notion of communalism, according to which the clergy should provide spiritual service, than 'the caesaropapist Ghibellinism of many German humanists.' This did not prevent him from close friendships with the humanists of Strasbourg, particularly with Jakob Wimpfeling, who enjoyed Geiler's support for his project of erecting a civic Latin school in the city. Education, however, was not exclusively a humanist preoccupation, and, other than noting the social proximity of humanists and their potential informal influence on Geiler's preaching style, the impact of humanist philological, textual, and rhetorical foci on his sermons is not of concern here.[84]

This section has described the intellectual and theological trends and debates of Catholicism in the late fifteenth century, within which Geiler was educated and composed his sermons. He is identified as subscribing to ideas most commonly found within the *via moderna* and articulated views that emphasized God's voluntary commitment to Christians, such as the guarantee of his side of the sacramental agreement and the dependability of the natural world's continuance. Fallen angels were included by Geiler as active in nature and their influence was comfortably incorporated alongside the inherent features of God's design and divine intervention as acceptable explanations for natural events and processes in his sermons.

Assumptions about supernatural and divine intervention in the natural world were integral to Catholic representations of the environment in the early sixteenth century. As a highly educated cleric, Geiler's easy command of Aristotelian natural philosophy and doctrinal orthodoxies regarding demons and the devil reinforced cultural views of nature which were the basis of apotropaic ritual behaviour described earlier in this chapter. Geiler enjoyed an ability to communicate successfully, among other things, the intricacies of late medieval scholastic natural and philosophy and demonology to his audience; this contributed to his importance as an influential cultural agent and emphasized his leadership in reinforcing Alsatian views of the natural world.

Conclusions

By developing the cultural, ritual, and intellectual frames for Geiler's 1509 representations of nature, not only are late medieval views of nature made more accessible, but we gain greater insight into the responses of pre-Reform Alsatians to the uncertain and unreliable conditions of the Spörer Minimum. Each aspect of the frame merits consideration, as coping with the higher-than-average volume of destructive weather events touched on all aspects of life in the Upper Rhine Valley.

Although conforming to Scripture in the broadest sense, the sermons of this learned Catholic preacher capture early sixteenth-century assumptions about the natural world which were based on centuries of elaboration in natural philosophy and theology. While God was clearly supreme within Geiler's cosmology in that He designed, created, and vivified Creation, the clergyman felt confident in describing the process, expanding the Biblical foundation to preach that God deposited His wisdom into every living being and that out of all Creation, only humans failed to act according to that God-given wisdom. Geiler told his audiences that God's wisdom was in the hare, the eagle, and the bear, and implied it was also in the tree, the bee, the sheep, the wolf, and the rest of Creation. His exposition of the seven reasons for wolf attacks explained that while the source of most events in nature was, in fact, within the natural world itself, there were still two possible causes that were attributed to spiritual actors. The devil and his swarming clouds of demons could, and did, pervert God's design, both independently of human desire and to fulfil the same, and God Himself could create unnatural events or behaviour in order to chastise or correct humans. Like other elements of the natural world, the weather also could be perverted from its design by diabolic agents or used by God for His ineffable purposes. Only His direct intervention in nature's normal operations could be recognized as a miracle; the meddling of demons was obliged to take place within the limits of divinely established natural law. Otherwise, and unless corrupted by the devil or demons, the animals, plants, weather, earth, and other parts of nature simply fulfilled their divinely designed function – even if this included inflicting destruction upon human beings, as in the attacks of wolves or hailstorms.

God's presence in nature is a thread which weaves throughout Geiler's conceptualization of nature: first through the design and construction of the world and each being on it, then in the animation appropriate to each one, and finally, in the wisdom with which he endows each creature. The level of divine immanence in this religious worldview nevertheless allowed for a striking degree of independent action within the mundane operations of the natural world, whereby many natural occurrences could be explained through the normal operations of divine intention, without diabolic or divine intervention.

Along with God's hand in their design, His animating life energy, and His wisdom, another commonality Geiler ascribed to all creatures was their vulnerability to possession by the devil, a demon, or more benign spiritual actors, like angels or God Himself. While spirits were powerful actors, though, human beings were portrayed as potential catalysts for action in a spiritual community which Geiler describes as a strict hierarchy. In that hierarchy, not only was humanity ranked below God, men and women were also inferior to his saints and angels, the devil and his demons, and the divinely established laws of nature. In this spiritual hierarchy, similar in structure to late medieval feudal society in western Germany, inferior beings (humans) chose a superior with whom to exchange pacts of obligation and protection (God or the devil). Geiler preached that engaging spiritual powers to act on one's behalf, whether maliciously or for protection against such malice, demanded an active choice by the person, who was obliged to contact the spirit of choice and beg them for aid. If the human was loyal to God, the pact was reliable and the sacraments could be relied upon to activate God's protection from danger. The devil, on the other hand, was portrayed as an unreliable overlord, one whose assistance or protection would fail at key moments.

Late medieval Alsatians were instructed by Geiler and his religious colleagues that there were means and ways of exerting indirect control over the natural environment, whether through prayers, rogations, exorcisms, or the ringing of bells which invoked God's power on their behalf or through enacting illegitimate rituals to summon an unreliable devil or demon. These methods had developed theological legitimacy during centuries of scholastic labour, matching and fostering a deep current of popular belief in otherworldly potency over a nature world infused with spiritual life.

Another element in Geiler's representation of the natural world was the contrast it allowed him between humans, for whom salvation was available, and non-humans, for whom it was not. That is, according to late medieval European cosmology, ants and demons shared at least one common feature: they were not members of the Roman Catholic Church. Barring a dramatic reconciliation between God and His fallen angels, the insects and the diabolic alike were denied any possibility of attaining eternal salvation, the first due to their lack of immortal soul and the second because of their rebellion against God. Therefore, both groups were able to serve Geiler as an 'Other' against whom Christian standards were established and Christian behaviour measured.

Geiler told his audiences that ants, known in the medieval bestiary for their industriousness, prudence, and steadfastness in faith, were obedient to God and compliantly executed their design according to His wisdom in an orderly, hard-working fashion, and that they were a worthy model to emulate. However, demons and the devil were described as traitorously disobedient, perverting not only their own design but the design of other creatures, such as individual plants and animals, the earth, the sky, and the weather. Geiler upheld the former as a moral exemplar for humanity, and scorned the latter as disorderly bringers of chaos. Praiseworthy orderliness and contemptible disorderliness both reveal the centrality of order itself to Geiler's representation of the divinely created natural environment.

What can be known of God in light of Geiler's conceptual organization of nature? In other words, reading against the grain of his rhetoric, what theological views are we able to discern from the manner in which God relates to His creation as preached by Geiler? A central observation is that Geiler's God sits at the pinnacle of a spiritual community which is ranked according to power and ability. He closely resembles a feudal overlord, as mentioned earlier, one who is idealized in terms of absolute power and command. A close relationship with Him provided protection from danger, while distance rendered one's relationship with the Lord vulnerable to co-optation by competitors. There are close parallels between the late medieval feudal social hierarchy in western Germany, the spiritual hierarchy asserted by Geiler, and the hierarchy of nature encapsulated in the concept of the Great Chain of Being, which ranks creatures in the order of their perceived proximity to God.

While it is impossible to claim that Geiler's emphasis on hierarchy in the natural world correlates directly with his preference for a strict social hierarchy, it is equally impossible to deny the conceptual similarity between the static, ranked social order he advocated, and the divinely ordained hierarchy he assumed as the unquestioned structure of the natural environment. It is also possible to speculate that in 1509, Geiler was attempting to promote social stability by using his influence to deliver a view of the natural world as intentionally and contentedly hierarchically ordered, particularly in light of the *Bundschuh* rebellions and the social tensions in Alsace which were likely obvious to the preacher.

A second observation is that, according to Geiler, comprehension of God's natural order was available through humanity's intellectual abilities. His view of the environment as rationally ordered, and accessible to comprehension through the human capacity for rational thought, had, by 1509, a millennium-long anchor in Augustine of Hippo's *On the Trinity*. In it, the Bishop argued for a direct link between God, creation, human nature, and our ability to reason, which resulted in traces of God in the human process of reasoning. Using, as McGrath notes, essentially Aristotelian categories of causation, Thomas Aquinas expanded on this to postulate a fundamental divine 'presence' in the natural world as a consequence of the relationship between creator and created. In *Summa contra Gentiles*, he stated that meditation upon God's works enabled an admiration of God's wisdom and reflection upon it. Thusly, he reasoned, it was possible to infer God's wisdom from reflecting upon God's works. Geiler's description of God's wisdom as bestowed upon every creature was itself a logical expansion of Aquinas' assertion.[85]

If, as the model of socio-economic metabolism suggests, cultural representations of nature will find expression in social behaviour, communal rogations and exorcisms ought to be fruitful sites of inquiry into the late medieval Alsatian religious understanding of weather, and they do not disappoint. Both ceremonies were a functional operation of the conceptual hierarchy described above in that they appealed to a superior rank's responsibilities to protect a vassal or subject – rogations as a prophylactic and exorcisms as a remedy. The source of danger is diabolic, and, hence, outside of the divine order designed and instilled in the natural world; prayers in both instances plead with the ultimate authority (God) to protect or re-establish the natural order as it was intended to be.[86]

Reflecting Geiler's confidence that God reliably honours the performance of a sacrament with the bestowal of grace, weather rogations and exorcisms were founded on communal confidence that engaging in ritually approved behaviour, such as rogations and exorcisms, with an attitude of faith and humility – *facere quod in se ist* – would enlist beneficial spiritual powers on their behalf. Conversely, Geiler's scorn for demon-summoning rituals was explicitly because they were *not* dependable: the devil was so estranged from the created order that he did not reliably honour the spiritual relationship which he himself had established when he provided summoning rituals to malicious humans. The devil was a remarkably poor choice as a spiritual liege lord. The implicit parallel with unreliable terrestrial lords, should they fail to honour their responsibilities, was unlikely to have escaped Geiler's audience.

At least with respect to their relationship with weather and the natural world (because rogations were conducted to protect against other disasters besides weather, and many of nature's threats in addition to bad weather called for the cleansing power of an exorcism), late medieval Alsatians particurated in a ritual world shaped by the perspective of the *via moderna*. There was an expectation of full communal participation, signalled by the mention of fines for on-looking idlers and the *Rat*'s explicit request to preachers that they urge attendance; a stronger show of devotion by the community to the spiritual order exerted a more persuasive pressure on spiritual hierarchy, as it would upon the social hierarchy.

This case study presenting an orthodox Roman Catholic view of nature suggests that the root principles of late medieval Alsatian culture were the concept of hierarchy and the relationships which supported an orderly environment. When spiritual agents caused the natural world to threaten humanity, those threats sprang from outside agents, so to speak; the devil and his demons had rebelled against God's ordered Creation, and addressing those threats involved a plea to God, his angels, and/or the saints to nullify diabolic activity and re-establish God's orderly design. Reason was a valued tool in understanding this natural order; the sixteenth-century rational view, however, included accounting for powerful spiritual agents with the capacity to affect nature. This case study shows the manner in which nature itself – the geographical, biological, and atmospheric features which make up the natural environment – was represented by an orthodox Catholic preacher as a hierarchical society of divinely created material vessels, brought to life with a bolt of God's own energy, filled with His wisdom, and each with its own value in the *societatis rerum creatarum*.

This study of *Die Emeis* contributes to dispelling Max Weber's portrait of late medieval Catholic theology as entirely irrational and dominated by 'mysterious, incalculable forces' and magic, particularly in light of a seminal essay examining the relationship of rationality and magic by Joseph Agassi and Ian Charles Jarvie, 'The Problem of Rationality and Magic.' In it, the authors note that a person may act rationally, believe rationally, or both. If they act rationally upon the basis of irrational beliefs, such may be understood as rationality in a weak sense. If they act rationally upon the basis of rationally held beliefs, this may be understood as rationality in a strong sense. As Hanska notes, medieval people frequently displayed rationality in the strong sense in that they were willing to act rationally upon rationally held beliefs, particularly in relation to preventing disasters or alleviating their effects.[87]

Strong and weak rationality are mixed in Geiler's sermons, although weak rationality predominates. For example, his exposition on the seven causes of wolf attacks describes five reasons for a possible wolf attack that were inherent in its own created and enwisened nature and two supernatural ones; his description of the causes of various dreams include material contexts and spiritual agents; and among his solutions to magically induced storms is found legislation against human action alongside Scripturally-supported ritual behaviour. While he consistently spoke rationally upon the basis of his beliefs, and consistently advised rational behaviour on those same beliefs, only occasionally were those rationally held beliefs. For the most part, and not unexpectedly from an orthodox preacher, the beliefs propounded from the pulpit were based on Geiler's Catholic faith. By the late fifteenth and early sixteenth centuries, elaborate philosophical and theological structures had developed within the faith which could explain events, processes, and forces of the natural environment. Geiler's education and position, as well as cultural expectations of him as a Doctor of Theology, among other factors, gave him the opportunity to represent these understandings of nature to his audience, and he did not fail.

Weak rationality, however, is not the absence of rationality altogether. The presence of rationality, even weakly, in the 1509 Lenten sermons of Johann Geiler von Kaysersberg contributes to the growing discreditation of Weber's 'disenchantment of the world' hypothesis by showing the existence of rational thought and behaviour in religious preaching before the Protestant Reformation.

To conclude, despite demonstrable stress from the natural environment upon the people of Alsace and ensuing social instability during several decades of the Spörer Minimum, Strasbourg's famous preacher continued to uphold orthodox Roman Catholic doctrine about nature, including how it was created and what methods were available for influencing weather events. With continuing environmental stresses, along with heightened pressure from anti-clerical views, the role of print in revolutionizing communication networks, corruption among the clergy, and many other social and cultural factors, Alsatian society began to show more agitation and cultural confidence in Catholic dogma eroded.

Notes

1 Johannes Geiler von Kaysersberg, *Die Emeis*, ed. Johannes Pauli (Strasbourg: Johann Grüninger, 1517), fol. IV[v]. The passage comes from Proverbs 6:6: *vade ad formicam o piger et considera vias eius et disce sapintiam* (go to the ant, o lazy one, and consider his ways and learn wisdom).

2 Voltmer, Rita, *Wie der Wächter auf dem Turm: ein Prediger und seine Stadt Johannes Geiler von Kaysersberg (1445–1510) und Strassburg* (Trier: Porta Alba Verlag, 2005), 948, 978, 980, and 982.

3 This and the following paragraphs, unless otherwise indicated, are based on the extensive biography of Johannes Geiler found in Voltmer, *Wie der Wächter*, 132–65.

4 Charles Guillaume A. Schmidt, *Histoire Littéraire de L'Alsace a la fin du XVe et au commencement du XVIe iècle*, 2 vols (repr. Hildesheim: Georg Olms, 1966; France: Sandoz et Fischbacher, 1879), 371–2.

5 Ibid., 343; Philippe Dollinger, *La ville libre de Strasbourg au carrefour des courants de pensée du XVIe siècle: Humanisme et Réforme a Strasbourg*, Exhibition organized by the Strasbourg Archives, Library, and Museums at the Old Customs House from 5 May to 10 June 1973 (Strasbourg: Ville de Strasbourg, 1973), 27.

6 Schmidt, 346, 360, and 372; As well as the evidence of his friendship with members of the humanist community in Strasbourg, in the biography he wrote upon Geiler's death Jacob Wimpfeling noted the welcome Italian humanists such as Pico della Mirandola and Ficino found at the preacher's table. Jacob Wimpfeling, *Vita Ioannis Keyserspergii*, in Johannes Geiler von Kaysersberg, *Sermones et varii tractatus*, ed. Peter Wickgram (Strasbourg: Johann Grüninger, 1517), fol. CLVI[v]; however, Geiler did not read Greek or Hebrew, showed no indication of the humanist preoccupation with philology, and, while sharing with humanists an interest in education, preferred that students be trained with materials from the Church Fathers rather than classical poets. Rita Voltmer, *Wie der Wächter auf dem Turm: ein Prediger und seine Stadt Johannes Geiler von Kaysersberg (1445–1510) und Strassburg* (Trier: Porta Alba Verlag, 2005), 54; E. Jane Dempsey Douglass, *Justification in Late Medieval Preaching: A Study of John Geiler of Keisersberg* (Leiden: Brill, 1966), 206–7; Having been named Imperial Preacher in 1498, Geiler was summoned to Füssen in order to offer private counsel to Holy Roman Emperor Maximilian I in 1503. Schmidt, 368–9.

7 Biographies from Geiler's contemporaries include Jakob Wimpfeling, *In Iohannis Keiserspergii Theologi doctrina vitaque probatissimi primi Argentinensis ecclesiae praedicatoris mortem planctus et lamentation cum aliquali vitae suae description et quorundam epithaphiis* (Jakob Köbel, 1510) and Daniel Specklin, Daniel, *Les Collectanées de Daniel Specklin, chronique Strasbourgeoise du Seizième Siècle*, ed. Rodolphe Reuss, Fragments des Anciennes Chronique d'Alsace, II (Strasbourg: Librairie J. Noiriel, 1890). Strasbourg's first municipal architect, Specklin wrote a two-volume chronicle about his city; only fragments survive. Noteworthy biographers of Geiler during the nineteenth century include Léon Dacheux, *Un réformateur Catholique à la Fin du XVe siècle Jean Geiler de Kaysersberg, prédicateur à la cathedrale de Strasbourg: 1478–1510; étude sur sa vie et son temps* (Paris: Delagrave; Strasbourg: Derivaux, 1876) and Charles Guillaume A. Schmidt, *Histoire Littéraire de l'Alsace a la fin du XVe et au commencement du XVIe iècle, 2 vols (France: Sandoz et Fischbacher, 1879; repr. Hildesheim: Georg Olms, 1966)*. In the twentieth century, significant research about Geiler was conducted by E. Jane Dempsey Douglass, *Justification in Late Medieval Preaching: A Study of John Geiler of Keisersberg* (Leiden: Brill, 1966), and Francis Rapp, *Réforme et Réformation à Strasbourg: Eglise et Société dans le Diocèse de Strasbourg (1450–1525)* (Paris: Ophrys, 1974). The twenty-first century has seen Rita Voltmer release an authoritative survey of Geiler's life and work, *Wie der Wächter auf dem Turm*; Voltmer, *Wie der Wächter*, 50–1.

8 Gerhard Bauer, 'Johannes Geiler von Kaysersberg (1445–1510) und seine Hexenpredigten in der Ameise,' in *Stregoneira e streghe nell'Europa moderna*, eds. Giovanna Bosco and Patrizia Castelli (Pisa and Rome: Pacini, 1996), 133–67; Rita Voltmer, 'Du discours à l'allégorie: Représentations de la superstition, de la magie et de la sorcellerie dans les sermons de Johannes Geiler de Kaysersberg,' in *Sorcellerie Savante et Mentalités Populaires*, eds. Antoine Follain and Maryse Simon (Strasbourg: Presses universitaires de Strasbourg, 2013), 46–52; Rita Voltmer, 'Preaching on Witchcraft: The Sermons of Johannes Geiler of Kaysersberg,' in *Contesting Orthodoxy in Medieval and Early Modern Europe*, eds. Louise N. Kallestrop and Raisa M. Toivo, Palgrave Historical Studies in Witchcraft and Magic series (Palgrave Macmillan, 2017), 192–215; Uwe Israel, *Johannes Geiler von Kaysersberg (1445–1510). Der Strassburger Münsterprediger als Rechtsreformer*, Berliner historische Studien, 27 (Berlin: Duncker & Humblot, 1997); Irmgard Weithase, 'Geiler von Kaisersberg als Kanzelredner,' *Sprechkunde und Sprecherziehung*, 4 (1959), 116–33; Julia Burkhardt, 'Learning from Bees, Wasps, and Ants. Communal Norms, Social Practices, and Contingencies of Nature in Medieval Insect Allegories,' in *Fragmented Nature: Medieval Latinate Reasoning on the Natural World and Its Order*, eds. Mattia Cipriani and Nicola Polloni, Studies in Medieval History and Culture (Abingdon: Routledge, Taylor & Francis Group, 2022); Rita Voltmer, 'Political Preaching and a Design of Urban Reform: Johannes Geiler of Kaysersberg and Strasbourg,' *Franciscan Studies*, 71 (2013), 71–87.
9 Tambora, www.tambora.org
10 Specklin, 461, Op. Cit. 2129 (Exc. Sp). As a comparison, it may be helpful to know that, on average, twentieth-century cherry trees in Alsace came to full bloom during the latter half of April. Anonymous author, *Ein geschrieben Buch oder Chronik von der Statt Straßburg: Item auch etwas von Augsburg und Baiern.* – Stadtarchiv Augsburg Nr. 55, Teil 1 und 2, year 1473; Sebastian Franck, *Chronica, Zeytbuch vnd geschychtbibel von anbegyn biß inn diß gegenwertig MDXXXI. jar : Darin beide Gottes vnd der welt lauff, hendel, art, wort, werck, thun, lassen, kriegen, wesen vnd leben ersehen vnd begriffen wirt...* (Straßburg: Balthassar Beck, 1531), Fol. CCX^r.
11 Niklaus Gerung, *Des Kaplans Niklaus Gerun genannt Blauenstein Fortsetzung des Flores Temporum 1417–1475*, in Oliver Wetter and Christian Pfister, 'Spring-Summer Temperatures Reconstructed for Northern Switzerland and Southwestern Germany from Winter Rye Harvest Dates, 1454–1970,' *Climate of the Past*, 7 (2011), 1318; Gabriela Signori, 'Ritual und Ereignis: Die Straßburger Bittgänge zue Zeit der Burgunderkriege (1474–1477),' *Historische Zeitschrift* 264 (April 1997), 282; Luzien Pfleger, 'Die Stadt- und Rats-Gottesdienste im Strassburger Münster,' *Archiv für Elsässische Kirchengeschichte* (1937), 32.
12 Sample provided by a survey of 1473 on the database www.Tambora.org, <www.tambora.org> [accessed 14 July 2022].; Jehan Aubrion, *Journal de Jehan Aubrion, bourgeois de Metz, avec sa continuation par Pierre Aubrion, 1465–1512*, ed. Lorédann Larchey (Metz: F. Blanc, 1857), 50–1. Note that during the twentieth century, wild cherries in Lorraine ripen on average in June, and the same stage of grape development tends to arrive in early to mid-July. *Letopis* are chronicles from medieval Slavic regions; the *Ustyug Letopis* record events from the city of Ustyug (today known as Veliky Ustyug (*Great Ustyug*)), located at the confluence of the Sukhona and Yug Rivers. Jon Henly, 'Hunger Stones, Wrecks and Bones: Europe's Drought Brings Past to the Surface,' *The Guardian*, 19 August 2022 <https://www.theguardian.com/world/2022/aug/19/hunger-stones-wrecks-and-bones-europe-drought-brings-past-to-surface> [accessed 22 August 2022]; Vyacheslav N. Razuvaev and others, *Extreme Heat Wave over European Russia in Summer 2010: Anomaly or a Manifestation of Climatic*

Trend?,' poster presentation at Global Environmental Change AGU Focus Group annual fall meeting, session entitled *Bringing Together Environmental, Socio-Economic and Climate Change Studies in Northern Eurasia*, San Francisco, USA, 15 December 2010; also available on ResearchGate <https://www.researchgate.net/publication/253866178_Extreme_Heat_Wave_over_European_Russia_in_Summer_2010_Anomaly_or_a_Manifestation_of_Climatic_Trend> [accessed 22 July 2022].

13 Surveys for 1474 and 1475 on the database www.Tambora.org, <www.tambora.org> [accessed 22 July 2022]; Franck, Fols. CCX^{r-v}; Karl Müller, *Geschichte des badischen Weinbaus. Mit einer badischen Weinchronik und einer Darstellung der Klimaschwankungen im letzten Jahrtausend* (Lahr in Baden: Schauenburg, 1953).

14 Surveys for 1476, 1477, 1478, and 1479 on the database www.Tambora.org, <www.tambora.org> [accessed 22 July 2022]; Anonymous, *Ein geschrieben Buch*, year 1476.

15 Survey for 1480 on the database www.Tambora.org, <www.tambora.org> [accessed 6 August 2022]; Johann Jacob Meyer, *La Chronique Strasbourgeoise*, ed. Rodolphe Reuss (Strasbourg: J. Noiriel, 1873), 122; Specklin, 466, Op. Cit. 2150 (Pp. Silb); Jean Wencker, 'Chronik und Zeitregister der Statt Strassburg,' in *Les Chroniques Strasbourgeoises de Jacques Trausch et de Jean Wencker*, ed. Léon Dacheux, Fragments des Anciennes Chroniques d'Alsace, III (Strasbourg: Imprimerie Strasbourgeoise and R. Schultz, 1892), 260, n2977; Peter Blickle, *The Revolution of 1525: The German Peasants' War from a New Perspective*, trans. By Thomas A. Brady, Jr. and H.C. Erik Midelfort (Baltimore and London: The Johns Hopkins University Press, 1985), 72; L. Pfleger, 35–6; Signori, 318.

16 Survey for 1481, 1482, and 1483 on the database www.Tambora.org, <www.tambora.org> [accessed 9 August 2022]; Wetter and Pfister, 50, fig. 4; Francis Rapp, The Social and Economic Prehistory of the Peasants' War in Lower Alsace,' *The German Peasant War of 1525 – New Viewpoints*, eds. Bob Scribner and Gerhard Benecke (London: George Allen & Unwin, 1979), 59; Strasbourg, Archives de la Ville et de l'Eurométropole de Strasbourg (AVES), Mandat et Réglements 1 MR 2, 115; Meyer, 122; Specklin, 467, Op. Cit. 2155 (K.-S. G).

17 Survey for 1484, 1485, 1486, 1487, 1488, 1489, and 1490 on the database www.Tambora.org, <www.tambora.org> [accessed 10 August 2022]; Strasbourg AVES, 1 MR 2, 118; Anonymous, *Ein geschrieben Buch*, year 1490; L. Pfleger, 37.

18 Survey for 1491 and 1492 on the database www.Tambora.org, <www.tambora.org> [accessed 11 August 2022]; Franz-Josef Mone, ed., *Quellensammlung der badischen Landesgeschichte*, Vols 1 and 2 (Karlsruhe: C. Macklot, 1848), Vol. 1, 300, Op. Cit. 86; Blickle, *The Revolution of 1525*, 72; Georges Bischoff, *La guerre des Paysans: L'Alsace et la révolution du Bundschuh 1493- 1525* (Strasbourg: La Nuée Blueu, 2010), 69 and 93; Andrew Cunningham and Ole Peter Grell, *The Four Horsemen of the Apolocalypse: Religion, War, Famine and Death in Reformation Europe* (Cambridge: Cambridge University Press, 2000), 15.

19 Bischoff, 93.

20 Ibid., 83–94.

21 Ibid., 96–7.

22 Survey for 1493 and 1494 on the database www.Tambora.org, <www.tambora.org> [accessed 22 August 2022]; Bischoff, 69.

23 Survey for 1495, 1496, 1497, 1498, and 1499 on the database www.Tambora.org, <www.tambora.org> [accessed 22–26 August 2022]; Meyer, 123; Specklin, p. 475, Op. Cit. 2176 *(Pp. Sch);* Heinrich Steinhöwel and Jakob Köbeln, *Beschreibunge einer Chronic Von anfang der Welt bisz auff Keyser Friderich den Dritten knrtz Summ irt vor Jam durch* (Franckfurt am Meyn: Christian Egenolff, 1531), [VD16 S 8811], fol. XLV^v.

24 Survey for 1500 and 1501 on the database www.Tambora.org, <www.tambora.org> [accessed 26 August 2022]; P.F. Malachias Tschamser, ed., *Annales oder Jahrs-Geschicthen der Baarfüferen oder Minderen Brüdern S. Franc. ord. inßgemein Conventualen gevannt, zu Thann*, 2 vols (Colmar: Hoffmann, 1864), vol. 1, 702.
25 Survey for 1502 on the database www.Tambora.org, <www.tambora.org> [accessed 30 August 2022]; Tschamser, I, 698; Bischoff, 101–4.
26 Bischoff, 104.
27 Ibid., 104.
28 Pierre Fluck, J. Gauthier, and A. Disser, 'The Alsatian Altenberg: A Seven Centuries Laboratory for Silver Metallurgy,' *Archéomètallurgie en Europe*, Troisième Conférence Internationale, France, 2011, available online at HAL (Hyper Articles en Ligne), https://hal.science/hal-00914305/document doi: hal- 00914305, version 1 [accessed 10 January 2024]; Personal communication with Pascal Hestin, deep-pit guide at Tellure Silver Mine museum, Ste-Marie-aux-Mines, 28 March 2013.
29 Survey for 1503 on the database www.Tambora.org, <www.tambora.org> [accessed 31 August 2022]; L. Pfleger, 38–9; Tschamser, Vol I, 702 and 704. See also Xavier Mossmann, ed., *Chronique des dominicains de Guebwiller* (Guebwiller: G. Brückert; Colmar: L. Reiffinger; Strasbourg: Schmidt et Grucker, 1844), 97–8.
30 Survey for 1504, 1505, and 1506 on the database www.Tambora.org, <www.tambora.org> [accessed 31 August 2022]; Meyer, 123; Tschamser, Vol. I, 706; Johannes Stedel, *Die Strassburger Chronik des Johannes Stedel*, ed. Paul Fritsch (Strasbourg: Sebastian Brant-Verl, 1934), 112; Meyer, 123; Franck, Fol. CCXIX^r; Tschamser, Vol. I, 710.
31 Survey for 1507, 1508, and 1509 on the database www.Tambora.org, <www.tambora.org> [accessed 31 August 2022]; Léon Dacheux, ed., *Fragments de Diverses Vieilles Chroniques*, Fragments des Anciennes Chroniques d'Alsace, IV (Strasbourg: Imprimerie Strasbourgeoise, 1901) 82, no. 3962; Specklin, 480, Op. Cit. 2197 (Exc. Sp); Tschamser, Vol. I, 712; Mossmann, 102.
32 Carol Symes, 'Out in the Open, in Arras: Sightlines, Soundscapes, and the Shaping of a Medieval Public Sphere,' in *Cities, Texts, and Social Networks, 400–1500: Experiences and Perceptions of Medieval Urban Space*, eds. Caroline J. Goodson, Anne E. Lester, and Carol Symes (Aldershot: Ashgate, 2010), 295–300.
33 François Joseph Böhm, *Description nouvelle de la Cathédrale de Strasbourg et de sa fameuse tour* (Strasbourg: J. Kürsner, 1743), 53–8; Alfred Pfleger, 'Wettersegen und Wetterschutz im Elsass,' *Archiv für Elsässische Kirchengeschichte*, 16 (1943), 268–9; Sebastien Brant, *Les Annales de Sébastien Brant*, ed. Léon Dacheux, Fragments des Anciennes Chroniques d'Alsace, III (Strasbourg: R. Schultz, 1892), 215.
34 A. Pfleger, 'Wettersegen und Wetterschutz,' 261.
35 Jussi Hanska, *Strategies of Sanity and Survival: Religious Responses to Natural Disasters in the Middle Ages*, Studia Finnica Historica 2 (Helsinki: Finnish Literature Society, 2002), 35; Lawrence M. Bryant, 'Configurations of the Community in Late Medieval Spectacles: Paris and London during the Dual Monarchy,' in *City and Spectacle in Medieval Europe*, eds. Barbara A. Hanawalt and Kathryn L. Reyerson (Minneapolis: University of Minnesota Press, 1994), 10.
36 Sicardus Cremonensis, *Sicardi Cremonensis episcopi Mitralis de officiis*, ed. Gábor Sarbak and Lorenz Weinrich (Turnhout: Brepols, 2008), col. 368; Guillaume Durand, *Rationale divinorum officiorum* (Lugduni: Apud haeredes Gulielmi Rouillii, sumptibus Petri Rousselet, 1612), Vol. 2, Book VI, Chaps. CI-CII, 392–95; Hanska, 55–6.
37 Jacob Twinger von Königshofen, 'Chronik des Jacob Twinger von Königshofen 1400 (1415),' in *Die Chroniken der oberrheinischen Städte. Straßburg*, eds. Karl

Gustav Theodor Schröder and Karl Hegel, Die Chroniken der deutschen Städte vom 14. bis ins 16. Jahrhundert, II (Leipzig: S. Hirzel, 1871), 773–4.

38 L. Pfleger, 46–9.
  L. Pfleger, 42, 46, and 50; Voltmer, *Wie der Wächter*, 509–10; Hanska, 35; Signori, 292.

39 Signori, 283.

40 Current street names are provided for ease of identification. L. Pfleger, 52; Philippe-André Grandidier, *Essais historique et topographiques sur l'église cathédrale de Strasbourg* (Strasbourg: Levrault, 1782), 382–4.

41 Grandidier, *Essaies*, 382; L. Pfleger, 35; Signori, 318.

42 Jean Lebeau and Jean Marie Valentin, eds., 'La commune d'Ostwald au Magistrat de Strasbourg [1510],' in *L'Alsace au siècle de la Réforme: 1482–1621: Textes et Documents* (Nancy: Presses universitaires de Nancy, 1985), 88.

43 Strassburg, Stadtarchiv, Serie V 136, 2. Papierhandschrift des 15. Jahrhunderts; see also A. Pfleger, 'Wettersegen und Wetterschutz,' p. 266.

44 A small note about the selections to follow: the volume published as *Die Emeis* includes sermons from both 1507 ('Herr der Kunig, ich diente gern,' a religious interpretation of a popular children's game) and 1509 (15 sermons as 'Die Emeis,' with the fifteenth sermon on the ants expanded into 25 more gathered under the title 'Von den Hexen und Unholden'). This research uses sermon material only from 'Die Emeis' and 'Von den Hexen und Unholde,' all delivered during Lent 1509.

45 First appearing in a ninth-century collection of canons, *Canon Episcopi* was a clause circulated through the *Decretum Gratiani* (causa 26, quaestio 5, canon 12) which asserted material flight to be impossible and an illusion imposed by the devil; belief in the material reality of flight was declared a heresy. Voltmer, 'Du discours à l'allégorie'; Voltmer, 'Preaching on Witchcraft?' 183-215.

46 Geiler, *Die Emeis*, fol. XXXIV^r and fols. XXXIV^v-XXXV^r. The *Quadriga*, or the 'fourfold sense of Scripture' was the most common manner of scriptural interpretation of the period, as follows: (1) the *literal* sense of Scripture, in which the text was taken at face value; (2) The *allegorical* sense, which interpreted certain otherwise obscure passages of Scripture to produce statements of doctrine; (3) the *tropological* or *moral* sense, which interpreted such passages to produce ethical guidelines for Christian conduct; and (4) the *anagogical* sense, which interpreted passages to indicate the grounds of Christian hope, pointing toward the future fulfilment of the divine promises in the New Jerusalem.

47 Hans Peter Broedel, *The "Malleus Maleficarum" and the Construction of Witchcraft: Theology and Popular Belief* (Manchester: Manchester University Press, 2003), 19; Bauer, 148–51.

48 Johannes Nider, *Incipit prologus formicarii* (Köln: 1473), Libri Secundi, Capitulum Secundum (87).

49 Geiler, *Die Emeis*, fols. X^v–XII^v; Ibid., fol. XIII^v.

50 Geiler's assumption that lions are solitary is most likely based on encounters or descriptions of captive animals, although he may be drawing parallels with the Eurasian lynx or Caspian lion, both of which tend towards solitary hunting.

51 Geiler, *Die Emeis*, fols. VIII^{r-v}.

52 Nider, Libri Primus, Capitulum Tertium (32–8).

53 Bischoff, 324; Fernand Braudel, *Civilization and Capitalism, 15th–18th Century: The Structures of Everyday Life*, trans. By Siân Reynolds (University of California Press, 1981), 387; Bischoff, 80 and 325; Jack Kelly, *Gunpowder: Alchemy, Bombards, and Pyrotechnics: The History of the Explosive That Changed the World* (New York: Basic Books, 2009), 64.

54 Invocavit is the first Sunday after Ash Wednesday and, as such, the first Sunday of Lent. Geiler, *Die Emeis*, fols. XVIII^v–XIX^r.
55 Edward P. Mahoney, 'Lovejoy and the Hierarchy of Being,' *Journal of the History of Ideas*, 48 (1987), 211–30.
56 Geiler, *Die Emeis*, fols. XXXXI^v–XXXXII^v and fols. XXXV^v–XXXVIII^r.
57 Geiler, *Die Emeis*, fol. XXXXV^r.
58 Note that the first known ascription of the use of human-shaped figures (dolls or manikins) for the purpose of malicious magic in Europe is found in the 1487 *Malleus Maleficarum*. Geiler, fol. XXXXV^v and fols. XXXIX^v–XXXIX^v. Mispagination on the part of Johannes Grüninger, Pauli's publisher, led to two adjacent folios sharing the same number (XXXIX). Geiler, *Die Emeis*, fol. LVI^r.
59 Geiler, *Die Emeis*, fol. XXXXIV^r and fol. XXXXIV^v.
60 Geiler, *Die Emeis*, fol. XXXXIII^r.
61 Geiler, *Die Emeis*, fol. XXXXII^r and fol. LV^v.
62 Geiler, *Die Emeis*, fols. LV^{r-v}.
63 Sarah Ferber, *Demonic Possession and Exorcism in Early Modern France* (London: Routledge, 2004), 17–21.
64 In 1184, Pope Lucius III issued the decretal *Ad abolendam*; in 1231, Pope Gregory IX issued the decretal *Ille humani generis*; in 1258, Pope Alexander IV issued the decretal *Quod super nonnullis*, and in 1326, Pope John XXII issues the bull *Super illius specula*; Michael D. Bailey, 'The Age of Magicians: Periodization in the History of European Magic,' *Magic, Ritual, and Witchcraft, 3* (2008), 9–15.
65 Valais, Lausanne, Vevey, Neuchâtel, Bern, Frieburg im Breisgau, and Basel. Edward Peters, 'Superstition, Magic and Witchcraft on the Eve of the Reformation,' in *Witchcraft and Magic in Europe: The Middle Ages*, eds. Karen Jolly, Catharina Raudvere, and Edward Peters (London: The Athlone Press, 2002), 238–45; Wolfgang Behringer, 'Climate Change and Witch-Hunting: The Impact of the Little Ice Age on Mentalities,' *Climate Change* 43 (1999), 335–51; Voltmer, 'Political Preaching,' 80; for a detailed discussion of the issue, see Voltmer, *Wie Der Wachter*, 51–69.
66 These works included *Physics, De Caelo, On Generation and Corruption, On the Soul, Meteorology, Parva Naturalia*, and biological works such as *The History of Animals, The Parts of Animals*, and the *Generation of Animals*. Edward Grant, *The Nature of Natural Philosophy in the Late Middle Ages*, (Studies in Philosophy and the History of Philosophy, Volume 52 (Washington: The Catholic University of America Press, 2010), 32. I am grateful to this work and the following for informing much of what follows in this section: Sachiko Kusukawa, *The Transformation of Natural Philosophy: The Case of Philip Melanchthon* (Cambridge: Cambridge University Press, 1995), 1–27; Alistair E. McGrath, *The Intellectual Origins of the European Reformation*, 2nd Edition (Oxford: Blackwell Publishing, 2004), 67–88; William J. Courtenay, 'Antiqui and Moderni in Late Medieval Thought,' *Journal of the History of Ideas*, 48 (1987), 3–10; Heiko A. Oberman, 'Via Antiqua and Via Moderna: Late Medieval Prolegomena to Early Reformation Thought,' *Journal of the History of Ideas*, 48 (1987), 23–40; Norman Kretzmann, Anthony Kenny, and Jan Pinborg, eds., *The Cambridge History of Later Medieval Philosophy* (Cambridge: Cambridge University Press, 1982).
67 Grant, 51–3.
68 Grant, 308.
69 Grant, 45–6; "It is no exaggeration to characterize medieval Aristotelianism as empiricism without observation." Ibid., 221.

70 Grant, 223; McGrath, 68; Courtenay, 7.
71 McGrath, 80–2.
72 McGrath, 356; Courtenay, 7.
73 Oberman, 24
74 Hans-Jochen Schiewer, 'Universities and Vernacular Preaching: The Case of Vienna, Heidelberg, and Basle,' in *Medieval Sermons and Society: Cloister, City, University*, Proceedings of International Symposia at Kalamazoo and New York, eds. Jacqueline Hamesse, Beverly Mayne Kienzle, Debra L. Stoudt, and Anne T. Thayer, Fédération Internationales des Instituts d'Études Médiévales, Textes et Études du Moyen Âge, IX (Louvain-La-Neuve: Collège du Cardinal Mercier, 1998), 396.
75 Schmidt, 5.
76 While Voltmer (*Wie der Wächter*, 216–7) perceives Geiler as engaged in an imaginary exchange with his audience, there are suggestions that late medieval Catholic sermon audiences were not necessarily quiet, orderly, or passive, and Geiler may have as easily been responding to heckling or questions from the audience as imagining their occurrence.
77 Brady, Protestant Politics, 21; Geiler speaks of sin as 'negative space,' where it is a "privation of the splendor of the soul like a shadow which is a privation of light", Johannes Geiler, 'De xii fructibus spiritus sancti' in *Sermones prestantissimi ... fructuosissimi de tempore et de sanctis accomodandi (Argentine, 1515); Sed ais: quid de puero enutrito in nemore et solitudine aut pagano nunquid huius ligula cordis: tendit in deum per fidem? Respondeo post beatum Thomam in 3. scripti. Si facit quod in se est: utendo donis suis connatis et naturalibus. illuminabit wel per se wel per angelum aut hominem. Si hoc non fecerit sed cor suum peccatis infecerit: abutens naturalibus donis: non est mirum si non attrahitur per magnetem illum...* Johannes Geiler, *Nave penitentie*, ed. Johannes Otther (Augsberg: Augusta vindelicorum, 1511), fol. XIPv; Douglass, 112–20.
78 Douglass, 104; Geiler, *Die Emeis*, fol. XXVIII^r.
79 Aristotle, *Metereologica*, trans. By H.D.P. Lee (Cambridge: Harvard University Press, 1952), 68–71 and 222–31; Geiler, *Die Emeis*, fols. XXXXI^v–XXXXII^v; Stuart Clark, *Thinking with Demons: The Idea of Witchcraft in Early Modern Europe* (Oxford: Oxford University Press, 1997).
80 Thomas Aquinas, *Compendium theologiae*, transl. Cyril Vollert, S.J. (St. Louis & London: B. Herder Book Co., 1947), lib. 1, ch. 136.
81 Clark, 153.
82 For a fuller discussion of these opinions, see Voltmer, 'Du discours à l'allégorie,' 65–70.
83 McGrath, 40; Thomas A. Brady, Jr., '"You Hate Us Priests:" Anticlericalism, Communalism, and the Control of Women at Strasbourg in the Age of the Reformation,' in *Anticlericalism in Late Medieval and Early Modern Europe*, eds. Peter A. Dykema and Heiko A. Oberman (Leiden: Brill, 1993), 183; This support was articulated in Geiler's letter to Wimpfeling. Johannes Geiler von Kaysersberg, *Die ältesten Schriften Geilers von Kaysersberg*, ed. Léon Dacheux (Amsterdam: Rodopi, 1965; repr. Freiburg im Breisgau: Herder, 1882), 101–2.
84 McGrath, 159–62; Aquinas, II.ii. 3–4.
85 See also Catherine Bell, 'Constructing Ritual,' *Readings in Ritual Studies*, ed. Ronald Grimes (New Jersey: Prentice Hall, 1996), 21–33.
86 Ian Charles Jarvie and Joseph Agassi, 'The Problem of Rationality and Magic,' in *Rationality*, ed. Bryan Wilson (New York: Harper & Row, 1970), 173–93; Hanska, 46.

References

Primary sources

Anonymous, *Ein geschrieben Buch oder Chronik von der Statt Straßburg: Item auch etwas von Augsburg und Baiern*, Stadtarchiv Augsburg Nr. 55, Teil 1 und 2.

Aquinas, Thomas, *Compendium theologiae*, transl. Cyril Vollert, S.J. (St. Louis & London: B. Herder Book Co., 1947).

Archives de la Ville et de l'Eurométropole de Strasbourg (AVES), Strasbourg, *Mandat et Réglements* 1 MR 2.

Aristotle, *Metereologica*, trans. By H.D.P. Lee (Cambridge: Harvard University Press, 1952).

Aubrion, Jehan, *Journal de Jehan Aubrion, bourgeois de Metz, avec sa continuation par Pierre Aubrion, 1465–1512*, ed. Lorédann Larchey (Metz: F. Blanc, 1857).

Brant, Sebastien, *Les Annales de Sébastien Brant*, ed. Léon Dacheux, Fragments des Anciennes Chroniques d'Alsace, III (Strasbourg: R. Schultz, 1892).

Cremonensis, Sicardus, *Sicardi Cremonensis episcopi Mitralis de officiis*, eds. Gábor Sarbak and Lorenz Weinrich (Turnhout: Brepols, 2008).

Dollinger, Phillippe, *La ville libre de Strasbourg au carrefour des courants de pensée du XVIe siècle: Humanisme et Réforme a Strasbourg*, Exhibition organized by the Strasbourg Archives, Library, and Museums at the Old Customs House from 5 May to 10 June 1973 (Strasbourg: Ville de Strasbourg, 1973).

Durand, Guillaume, *Rationale divinorum officiorum* (Lugduni: Apud haeredes Gulielmi Rouillii, sumptibus Petri Rousselet, 1612).

Franck, Sebastian, *Chronica, Zeytbuch vnd geschychthibel von anhegyn biß inn diß gegenwertig MDXXXI. jar : Darin beide Gottes vnd der welt lauff, hendel, art, wort, werck, thun, lassen, kriegen, wesen vnd leben ersehen vnd begriffen wirt...* (Straßburg: Balthassar Beck, 1531).

Geiler von Kaysersberg, Johannes, 'De xii fructibus spiritus sancti' in *Sermones prestantissimi ... fructuosissimi de tempore et de sanctis accomodandi (Argentine, 1515); Sed ais: quid de puero enutrito in nemore et solitudine aut pagano nunquid huius ligula cordis: tendit in deum per fidem? Respondeo post beatum Thomam in 3. scripti. Si facit quod in se est: utendo donis suis connatis et naturalibus. illuminabit wel per se wel per angelum aut hominem. Si hoc non fecerit sed cor suum peccatis infecerit: abutens naturalibus donis: non est mirum si non attrahitur per magnetem illum...* Johannes Geiler, *Nave penitentie*, ed. Johannes Otther (Augsberg: Augusta vindelicorum, 1511).

———, *Die Emeis*, ed. Johannes Pauli (Strasbourg: Johann Grüninger, 1517).

——— *Die ältesten Schriften Geilers von Kaysersberg*, ed. Léon Dacheux (Freiburg im Breisgau: Herder, 1882; repr. Amsterdam: Rodopi, 1965).

Grandidier, Philippe-André, *Essais historique et topographiques sur l'église cathédrale de Strasbourg* (Strasbourg: Levrault, 1782).

Lebeau, Jean, and Jean Marie Valentin, ed., 'La commune d'Ostwald au Magistrat de Strasbourg [1510],' in *L'Alsace au siècle de la Réforme: 1482–1621: Textes et Documents* (Nancy: Presses universitaires de Nancy, 1985).

Meyer, Johann Jacob, *La Chronique Strasbourgeoise*, ed. Rodolphe Reuss (Strasbourg: J. Noiriel, 1873).

Mossmann, Xavier, ed., *Chronique des dominicains de Guebwiller* (Guebwiller: G. Brückert; Colmar: L. Reiffinger; Strasbourg: Schmidt et Grucker, 1844).

Nider, Johannes, *Incipit prologus formicarii* (Köln: 1473).

Specklin, Daniel, *Les Collectanées de Daniel Specklin, chronique Strasbourgeoise du Seizième Siècle*, ed. Rodolphe Reuss, Fragments des Anciennes Chronique d'Alsace, II (Strasbourg: Librairie J. Noiriel, 1890).

Stadtarchiv, Strassbourg, Serie V 136, 2. Papierhandschrift des 15. Jahrhunderts.

Stedel, Johannes, *Die Strassburger Chronik des Johannes Stedel*, ed. Paul Fritsch (Strasbourg: Sebastian Brant-Verl, 1934).

Steinhöwel, Heinrich, and Jakob Köbeln, *Beschreibunge einer Chronic Von anfang der Welt bisz auff Keyser Friderich den Dritten knrtz Summ irt vor Jam durch ... Hern Heinrichen Steinhöwel Doctorn Stattartzt zů Vlm auszgezogen vnnd gemacht. Vnnd ietzo durch den Erfarnen H. Jacob Koebeln Stattschreiber zů Oppenheym an etlichen Ortenn gemeret vnd auff Keyser Carlen den V. erstreckt. Mit anhang beschreibung der zeit Jsidori* (Franckfurt am Meyn: Christian Egenolff, 1531).

Stolz, Hans, *Die Hans Stolz'sche Gebweiler Chronik. Zeugenbericht über den Bauernkrieg am Oberrhein*, ed. Wolfram Stolz (Freiburg: Edition Stolz, 1979).

Tschamser, P.F. Malachias, ed., *Annales oder Jahrs-Geschicthen der Baarfüferen oder Minderen Brüdern S. Franc. ord. inßgemein Conventualen gevannt, zu Thann*, 2 vols (Colmar: Hoffmann, 1864).

Twinger von Königshofen, Jacob, 'Chronik des Jacob Twinger von Königshofen 1400 (1415),' in *Die Chroniken der oberrheinischen Städte. Straßburg*, eds. Karl Gustav Theodor Schröder and Karl Hegel, Die Chroniken der deutschen Städte vom 14. bis ins 16. Jahrhundert, II (Leipzig: S. Hirzel, 1871).

Wencker, Jean, 'Chronik und Zeitregister der Statt Strassburg,' in *Les Chroniques Strasbourgeoises de Jacques Trausch et de Jean Wencker*, ed. Léon Dacheux, Fragments des Anciennes Chroniques d'Alsace, III (Strasbourg: Imprimerie Strasbourgeoise and R. Schultz, 1892).

Wimpfeling, Jacob, 'Vita Ioannis Keyserspergii,' in *Johannes Geiler von Kaysersberg, Sermones et varii tractatus*, ed. Peter Wickgram (Strasbourg: Johann Grüninger, 1517).

Secondary sources

Bailey, Michael D., 'The Age of Magicians: Periodization in the History of European Magic,' *Magic, Ritual, and Witchcraft*, 3:1 (2008), 1–28.

Bauer, Gerhard, 'Johannes Geiler von Kaysersberg (1445–1510) und seine Hexenpredigten in der Ameise,' in *Stregoneira e streghe nell'Europa moderna*, eds. Giovanna Bosco and Patrizia Castelli (Pisa and Rome: Pacini, 1996).

Behringer, Wolfgang, 'Climate Change and Witch-Hunting: The Impact of the Little Ice Age on Mentalities,' *Climate Change* 43 (1999), 335–51.

Bell, Catherine, 'Constructing Ritual,' in *Readings in Ritual Studies*, ed. Ronald Grimes (New Jersey: Prentice Hall, 1996).

Bischoff, Georges, *La guerre des Paysans: L'Alsace et la révolution du Bundschuh 1493–1525* (Strasbourg: La Nuée Bleue, 2010).

Blickle, Peter, *The Revolution of 1525: The German Peasants' War from a New Perspective*, trans. by Thomas A. Brady, Jr., and H.C. Erik Midelfort (Baltimore and London: The Johns Hopkins University Press, 1985).

Böhm, François Joseph, *Description nouvelle de la Cathédrale de Strasbourg et de sa fameuse tour* (Strasbourg: J. Kürsner, 1743).

Brady, Thomas A., Jr., '"You Hate Us Priests:" Anticlericalism, Communalism, and the Control of Women at Strasbourg in the Age of the Reformation,' in *Anticlericalism in Late Medieval and Early Modern Europe*, eds. Peter A. Dykema and Heiko A. Oberman (Leiden: Brill, 1993).

—— *Protestant Politics: Jacob Sturm (1489–1553) and the German Reformation* (New Jersey: Humanities Press International, 1995).

Braudel, Fernand, *Civilization and Capitalism, 15th–18th Century: The Structures of Everyday Life*, trans. by Siân Reynolds (University of California Press, 1981).

Broedel Hans Peter, *The "Malleus Maleficarum" and the Construction of Witchcraft: Theology and Popular Belief* (Manchester: Manchester University Press, 2003).

Bryant, Lawrence M., 'Configurations of the Community in Late Medieval Spectacles: Paris and London during the Dual Monarchy,' in *City and Spectacle in Medieval Europe*, eds. Barbara A. Hanawalt and Kathryn L. Reyerson (Minneapolis: University of Minnesota Press, 1994).

Burkhardt, Julia, 'Learning from Bees, Wasps, and Ants. Communal Norms, Social Practices, and Contingencies of Nature in Medieval Insect Allegories,' in *Fragmented Nature: Medieval Latinate Reasoning on the Natural World and Its Order*, eds. Mattia Cipriani and Nicola Polloni, Studies in Medieval History and Culture (Abingdon: Routledge, Taylor & Francis Group, 2022).

Clark, Stuart, *Thinking with Demons: The Idea of Witchcraft in Early Modern Europe* (Oxford: Oxford University Press, 1997).

Courtenay, William J., '*Antiqui* and *Moderni* in Late Medieval Thought,' *Journal of the History of Ideas*, 48:1 (1987), 3–10.

Cunningham, Andrew, and Ole Peter Grell, *The Four Horsemen of the Apolocalypse: Religion, War, Famine and Death in Reformation Europe* (Cambridge: Cambridge University Press, 2000).

Dacheux, Léon, *Un réformateur Catholique à la Fin du XVe siècle Jean Geiler de Kaysersberg, prédicateur à la cathedrale de Strasbourg: 1478–1510; étude sur sa vie et son temps* (Paris: Delagrave; Strasbourg: Derivaux, 1876).

Douglass, E. Jane Dempsey, *Justification in Late Medieval Preaching: A Study of John Geiler of Keisersberg* (Leiden: Brill, 1966).

Ferber, Sarah, *Demonic Possession and Exorcism in Early Modern France* (London: Routledge, 2004).

Fluck, Pierre, J. Gauthier, and A. Disser, 'The Alsatian Altenberg: A Seven Centuries Laboratory for Silver Metallurgy,' in *Archéomètallurgie en Europe*, Troisième Conférence Internationale, France 2011, available online at HAL (Hyper Articles en Ligne), https://hal.science/hal-00914305/document, DOI: hal- 00914305, version 1 [accessed 10 January 2024].

Grant, Edward, *The Nature of Natural Philosophy in the Late Middle Ages*, Studies in Philosophy and the History of Philosophy, Volume 52, (Washington: The Catholic University of America Press, 2010).

Hanska, Jussi, *Strategies of Sanity and Survival: Religious Responses to Natural Disasters in the Middle Ages*, Studia Finnica Historica 2 (Helsinki: Finnish Literature Society, 2002).

Hestin, Pascal, deep-pit guide at Tellure Silver Mine museum, Ste-Marie-aux-Mines, France (personal communication), 28 March 2013.

Israel, Uwe, *Johannes Geiler von Kaysersberg (1445–1510). Der Strassburger Münsterprediger als Rechtsreformer*, Berliner historische Studien, 27 (Berlin: Duncker & Humblot, 1997).

Jarvie, Ian Charles, and Joseph Agassi, 'The Problem of Rationality and Magic,' in *Rationality*, ed. Bryan Wilson (New York: Harper & Row, 1970).

Kelly, Jack, *Gunpowder: Alchemy, Bombards, and Pyrotechnics: The History of the Explosive That Changed the World* (New York: Basic Books, 2009).

Kretzmann, Norman, Anthony Kenny, and Jan Pinborg, eds., *The Cambridge History of Later Medieval Philosophy* (Cambridge: Cambridge University Press, 1982).

Kusukawa, Sachiko, *The Transformation of Natural Philosophy: The Case of Philip Melanchthon* (Cambridge: Cambridge University Press, 1995).

Mahoney, Edward P., 'Lovejoy and the Hierarchy of Being,' *Journal of the History of Ideas*, 48:2 (1987), 211–30.

McGrath, Alistair, *The Intellectual Origins of the European Reformation*, 2nd Edition (Oxford: Blackwell Publishing, 2004).

Mone, Franz-Josef, ed., *Quellensammlung der badischen Landesgeschichte*, Vols 1 and 2 (Karlsruhe: C. Macklot, 1848).

Müller, Karl, *Geschichte des badischen Weinbaus. Mit einer badischen Weinchronik und einer Darstellung der Klimaschwankungen im letzten Jahrtausend* (Lahr in Baden: Schauenburg, 1953).

Oberman, Heiko A., 'Via Antiqua and Via Moderna: Late Medieval Prolegomena to Early Reformation Thought,' *Journal of the History of Ideas*, 48:1 (1987), 23–40.

Peters, Edward, 'Superstition, Magic and Witchcraft on the Eve of the Reformation,' in *Witchcraft and Magic in Europe: The Middle Ages*, eds. Karen Jolly, Catharina Raudvere, and Edward Peters (London: The Athlone Press, 2002).

Pfleger, Alfred, 'Wettersegen und Wetterschutz im Elsass,' *Archiv für Elsässische Kirchengeschichte*, 16 (1943), 259–72.

Pfleger, Luzien, 'Die Stadt- und Rats-Gottesdienste im Strassburger Münster,' *Archiv für Elsässische Kirchengeschichte*, 12 (1937), 1–55.

Rapp, Francis, *Réforme et Réformation à Strasbourg: Eglise et Société dans le Diocèse de Strasbourg (1450–1525)* (Paris: Ophrys, 1974).

——— 'The Social and Economic Prehistory of the Peasants' War in Lower Alsace,' in *The German Peasant War of 1525 – New Viewpoints*, eds. Bob Scribner and Gerhard Benecke (London: George Allen & Unwin, 1979).

Razuvaev, Vyacheslav N., P.Y. Groisman, O. Bulygina, and I. Borzenkova, *Extreme Heat Wave over European Russia in Summer 2010: Anomaly or a Manifestation of Climatic Trend?*, poster presentation at Global Environmental Change AGU Focus Group annual fall meeting *Bringing Together Environmental, Socio-Economic and Climate Change Studies in Northern Eurasia*, San Francisco, USA, 15 December 2010; also available on ResearchGate <https://www.researchgate.net/publication/253866178_Extreme_Heat_Wave_over_European_Russia_in_Summer_2010_Anomaly_or_a_Manifestation_of_Climatic_Trend> [accessed 22 July 2022].

Schiewer, Hans-Jochen, 'Universities and Vernacular Preaching: The Case of Vienna, Heidelberg, and Basle,' in *Medieval Sermons and Society: Cloister, City, University*, Proceedings of International Symposia at Kalamazoo and New York, eds. Jacqueline Hamesse, Beverly Mayne Kienzle, Debra L. Stoudt, and Anne T. Thayer, Fédération Internationales des Instituts d'Études Médiévales, Textes et Études du Moyen Âge, IX (Louvain-La-Neuve: Collège du Cardinal Mercier, 1998).

Schmidt, Charles Guillaume A., *Histoire Littéraire de L'Alsace a la fin du XVe et au commencement du XVIe iècle*, 2 vols (France: Sandoz et Fischbacher, 1879; repr. Hildesheim: Georg Olms, 1966).

Signori, Gabriela, 'Ritual und Ereignis: Die Straßburger Bittgänge zue Zeit der Burgunderkriege (1474–1477),' *Historische Zeitschrift*, 264:2 (1997), 281–328.

Symes, Carol, 'Out in the Open, in Arras: Sightlines, Soundscapes, and the Shaping of a Medieval Public Sphere,' in *Cities, Texts, and Social Networks, 400–1500: Experiences and Perceptions of Medieval Urban Space*, eds. Caroline J. Goodson, Anne E. Lester, and Carol Symes (Aldershot: Ashgate, 2010).

Voltmer, Rita, *Wie der Wächter auf dem Turm: ein Prediger und seine Stadt Johannes Geiler von Kaysersberg (1445–1510) und Strassburg* (Trier: Porta Alba Verlag, 2005).

—— 'Political Preaching and a Design of Urban Reform: Johannes Geiler of Kayserberg and Strasbourg,' *Franciscan Studies*, 71 (2013), 71–87.

—— 'Du discours à l'allégorie: Représentations de la superstition, de la magie et de la sorcellerie dans les sermons de Johannes Geiler de Kaysersberg,' in *Sorcellerie Savante et Mentalités Populaires*, eds. Antoine Follain and Maryse Simon (Strasbourg: Presses universitaires de Strasbourg, 2013), 45–88.

—— 'Preaching on Witchcraft? The Sermons of Johannes Geiler of Kaysersberg (1445–1510),' in *Contesting Orthodoxy in Medieval and Early Modern Europe*, eds. L. Kallestrup and R. Toivo, Palgrave Historical Studies in Witchcraft and Magic (Cham: Palgrave Macmillan, 2017), 192–215. https://doi.org/10.1007/978-3-319-32385-5_10.

Weithase, Irmgard, 'Geiler von Kaisersberg als Kanzelredner,' *Sprechkunde und Sprecherziehung*, 4 (1959), 116–33.

Wetter, Oliver, and Christian Pfister, 'Spring-Summer Temperatures Reconstructed for Northern Switzerland and Southwestern Germany from Winter Rye Harvest Dates, 1454–1970,' *Climate of the Past*, 7 (2011), 1307–26.

Digital sources

Henly, Jon, 'Hunger Stones, Wrecks and Bones: Europe's Drought Brings Past to the Surface,' *The Guardian*, 19 August 2022 (https://www.theguardian.com/world/2022/aug/19/hunger-stones-wrecks-and-bones-europe-drought-brings-past-to-surface), accessed 22 August 2022.

Tambora, www.tambora.org.

3 The common man and the natural world from Lent 1509 to 1525 (Case Study 2)

Introduction

The second and third decades of the sixteenth century were a moment in history when the common man and woman in Alsace bestirred themselves and propelled events forward; the focus of this chapter, then, turns to them. Two sources for an Alsatian peasant representation of the natural environment will be introduced: Clemens Zyegler and his pamphlet *Ein fast schon büchlin in welche(n) yederman findet ein hellen und claren verstandt von dem leib und blůt Christi* (*Ein fast schon büchlin*) and the *XII Artikels* composed by peasants in Swabia and copied throughout southwestern Germany during the 1525 Peasants' War, including in Alsace. As was the case with Geiler, the primary purpose of these tracts was not the communication of a religious representation of nature; Clemens Zyegler hoped to inspire within his peers an acceptance of his interpretation of the Eucharist, and the *XII Artikels* were a platform of demands by peasants in rebellion. After introducing these documents, the chapter goes on to provide a weather report for the years between 1509 and 1525, including some social consequences of the weather; to offer arguments for a distinctly peasant culture in early sixteenth-century Alsace and southwestern Germany; to establish a peasant representation of the natural environment; and, in conclusion, to summarize the social life of the Pure Word of God in Alsace until 1525 and to examine the peasant view of nature.

First, though: what is a peasant? The answer is not as simple as may first appear, particularly for contemporary readers who may have Monty Python or *Lord of the Rings* as referents (as delightful as these are). For much of the early twentieth century, the term was understood in Western academia to refer simply to small-scale agricultural producers who served as a starting point of social development. That is, rural peasants were either defined as European-style hide-bound traditionalists who unconsciously evolved (and would *all* evolve) into urban industrial workers as technological advances changed society or as proto-capitalists for whom unregulated markets were essential to creating agribusinesses. The persistence of peasants and peasantries into the

DOI: 10.4324/9781003262411-4

twenty-first century defied these confident predictions – almost one-third of humans are peasants today – and invited further investigation.

During the 1960s and 1970s, the translation into English of Soviet agronomist Alexander V. Chayanov's *Peasant Farm Organisation* (originally published in Russian in 1925) inspired a new interest in peasant studies which challenged both perspectives. Eric Vanhaute's recent summary is succinct and informative:

> Rural anthropologist Eric Wolf and rural sociologist Theodor Shanin, among others, moved this debate beyond a-historical and dichotomist representations. The question is not whether peasants were naturally conservative, values-rational, safety-oriented investors in their land and labor, or whether they tended to be risk-taking, market-oriented maximizers. They were and are both.[1]

Since Shanin and Wolf, the definition of a peasant and peasantries has become even more complex as peasants' relationship to markets, ecology, gender-related work, family labour, patterns of production and consumption, and inheritance patters came into question. Including non-European peasantries has proven particularly valuable in challenging Eurocentric ideas. The issue of power relations or class relations is currently central to the discussion, because a salient characteristic of peasants is their obligation to produce surpluses for powerful agents beyond their community, not only for their own sustenance. It is the appearance of external powers which marks the threshold between food cultivators in general and peasants in particular. Fundamentally assuming relationships with overarching states (and ensuing taxation), in 2018 the United Nations adopted the following definition as part of the Declaration on the Rights Of Peasants (UNDROP); it suits the needs of this chapter as well.

> For the purposes of the present Declaration, a peasant is any person who engages or who seeks to engage alone, or in association with others or as a community, in small-scale agricultural production for subsistence and/or for the market, and who relies significantly, though not necessarily exclusively, on family or household labour and other non-monetized ways of organizing labour, and who has a special dependency on and attachment to the land.[2]

In his edited volume exploring European peasantries, Tom Scott notes that the emergence of peasantries in Europe was an achievement of the High Middle Ages, one which saw a final evolution from clan or tribal retinue whose members bore arms <u>and</u> tilled the soil (this had prevailed particularly among German-speaking peoples) to a settled society with a separation of functions or duties. Some first hints of this division are visible in the military reforms of Charlemagne who, in 807 and 808 CE, restricted military service to those who held at least three, then four, hides of land. The three-fold division of medieval European society into Estates was solidified in 1152 when Holy Roman Emperor Friedrich I, also known as Frederick Barbarossa, proclaimed a *Großer Reichslandfrieden* (Ger. Great Imperial Peace) in which peasants were banned from owning weapons at all.[3]

Clemens Zyegler was a gardener in 1522, the moment when knowledge of him enters the historical record through his signature on a lease with the monks for a plot of land in the Krutenau (an area southeast of the island of Strasbourg around the convent of Saint-Nicolas-aux-Ondes). With 600 members, including shepherds and day labourers, the gardeners' guild was the largest in Strasbourg, albeit one with low prestige. Likely around 35 years old, Zyegler's civil status at the time was that of *Schultheißenbürger*; he was married, with a younger brother named Jörg who was a tailor living on Steinstrasse in Wolfgang Capito's parish of *Jung Sankt Peter* (today Saint-Pierre-le-Jeune). Indicating a birth somewhere other than Strasbourg, the *stettmeister* named Clemens *unser burger* (our citizen) on 25 February 1525, declaring his status as a mayoral citizen (usually a poorer person who had purchased citizenship at a lower fee and with fewer rights); Zyegler obtained the right of full citizenship on 18 December 1525.[4]

Zyegler almost drowned in the floods of 1524, and interpreted his survival as having been miraculously spared in order to publicly proclaim the Gospel. He appears to have been a literate though uneducated man: he quoted the Old Testament from a pre-Lutheran translation and the New Testament from Luther's 1522 translation (printed in Strasbourg before that year concluded). Some of the gardener's later works may have been a response to Caspar Hedio's *Chronicle*, published in 1530. Most relevant for this research, Zyegler was the author of *Ein fast schon büchlin*, a pamphlet printed in Strasbourg during the early months of 1525, likely by Johann Schwann (a former Franciscan monk).[5]

Although he was not one of the military leaders of the rebellion, Zyegler is considered to have been one of the most visible and most important early radical preachers in Strasbourg. His early preaching and writings raised three issues which would prove highly divisive: economic justice, the Eucharist, and baptism. Scholar of Anabaptism John Derksen describes Zyegler as having 'the enthusiasm of a populist reformer,' combining spiritualism, socialism, pacifism, and a type of apocalyptic universalism with obedience to the Word, an emphasis on ethics, and impatience with partial reform. He differed from the Lutheran reformers most sharply on the social implications of the Gospel, arguing that since God's sovereignty eliminates human hierarchy, there should be an end to exploitation and a reversal of economic institutions, and suggesting that Christ would have been a *Bundschuher*.[6]

It's difficult to know whether Zyegler radicalized his fellow gardeners or whether they radicalized his views, since in 1523 – the year before his near-death experience – the gardeners of St. Aurelia refused to pay tithes and rents to their lords. By spring 1524, when a crowd of gardeners threatened those convents they considered most corrupt in the city, Zyegler was likely urging them on. In August that year, the gardeners of *Jung Sankt Peter* parish set out to clear their church of images and, in September, a crowd of 400 dragged the anti-reform Augustinian Konrad Treger before the *Rat*, smashing statues along the way. Clemens' brother Jörg is known to have been in the mob. In response to these actions, Strasbourg's *Rat* offered to reduce taxes, agreed to use the tithe to remunerate reformers, and hired mercenaries to safeguard the city (instead of guildsmen). They allowed some religious images to be removed, ordered the preachers to avoid controversy in their sermons, and censored divisive publications.[7]

These attempts by the *Rat* to maintain order through compromise bore little fruit. In November 1524, the gardeners of St. Aurelia demolished their church's crucifix and tomb, while those in *Jung Sankt Peter*'s parish stripped the altars. In December, more churches endured the desecration of hosts and the smashing of images, and in February 1525, gardeners and barrelmakers demolished the main altar of *Stephanskapelle* (St. Stephen's church). Along with this active destruction of Catholic material culture, Strasbourg's radicals withheld their tithes, refused infant baptism, criticized Lutheran reformers, and kept the whole law of the Gospel, including the Sabbath.[8]

Zyegler's pamphlets gain authority as a source for accurate peasant representations of the natural world when it is understood that shortly after the publication of *Ein fast schon büchlin*, he went to Obernai and Barr to preach the Gospel upon an invitation by members of the gardeners' guild there. There are records of Zyegler preaching in meadows between Börsch and Saint Leonard, on the outskirts of Bernhardswiller, in the cemetery at Heiligenstein, and possibly at the Haart River; unrecorded events or vanished records likely hide several more sermons. The peasants around these towns at the foot of Mont Ste Odile were, on 3 March 1525, among the first to rebel in Alsace, and they rose up against Strasbourg's cathedral chapter, whose land they farmed. On 16 April, Zyegler was chosen as one of three representatives of the peasants to negotiate with the council of Dorlisheim. When the Strasbourg gardeners joined the rebellion, several of them came to join this band in ransacking the abbey of Altorf. Zyegler, however, found in this last a betrayal of his pacifist ideals; he left the peasant army at that point and returned to Strasbourg, escaping the slaughter to come. Nevertheless, despite his personal absence, a striking indication of his influence during the German Peasants' War was the rallying cry and banner which inspired those particular peasants: they fought for 'Gospel, Christ, and Clemens Zyegler.' His occupation, his radical religious views, and his central role in the Peasants' War in Alsace unite to merit a central role for *Ein fast schon büchlin* in this chapter.[9]

The second item to be introduced for the case study is not a single document, but the peasants' *XII Artikels*, the list of demands developed by members of the Upper Swabian peasant army. Journeyman furrier Sebastian Lotzer and Christoph Schappeler, the evangelical pastor of St. Martin's church in Memmingen, summarized the list, gave it its general form, and supplemented it with supporting scriptural citations. Similar demands were first made in Alsace during the 1513 *Bundschuh* rebellion, when the anonymous author was presumably among Joss Fritz' rebels. These were quickly printed by some 20 printers in 15 different locations, and, whether adopted in whole, in part, or amended, provided conceptual unity to a rebellion which spanned several regions from Alsace to west Saxony, Styria, and the Tyrol. The claims were justified through a religious frame in which the natural world was God's gift to all mankind. While economic, social, and political analyses of these demands are valuable, including an analysis of their religious aspect contributes towards a cultural understanding of the peasant perception of the natural environment around them.[10]

Weather report from 1509 to 1525 and some social consequences of that weather

The very snowy winter of 1509 mentioned in Chapter 2 was accompanied by a severe cold snap (birds were reported to have fallen dead from the air), followed by a summer drought stretching as far from Strasbourg as Göttingen, Tübingen, and Rothenburg ob der Tauber. Water levels in Alsace dropped so much that some regional water mills were unable to function and the wine harvest was half the anticipated volume. The following year, on 1 April 1510 (*11 April*), the day following Misericordia (also known as the Third Sunday of Easter or the day that occurs two weeks after Easter Sunday), a rogation for good weather and a successful harvest was organized; it began from the *Liebfrauenmünster* of Strasbourg and concluded with a pilgrimage. Despite the collective plea to God, the Friday following Corpus Christi of 1510 (*31 May/10 June*) had heavy flooding that lifted bridges, uprooted trees, and killed 400 head of cattle in the Ried (the area between the Ill and Rhine Rivers from Strasbourg to Colmar). Despite the flood, the wine harvest of 1510 was abundant.[11]

The winter of 1511, however, was severely cold and damaged the Alsatian grape vines and fruiting trees, followed by a wet summer where high water levels, flooding, and property damage are recorded for the Rhine, Ill, Thur, and Large Rivers. Special weather rogations were organized by Strasbourg's *Rat* and *Hochstift*, but in vain, for the harvests of wine and agricultural products were failures.[12]

The regional weather during the summer of 1512 was generally rainy and dank once again, leading to a second year of flooding by the Rhine and a wine harvest that was reported to be scant (but tasty). Landlords pressed for higher taxes despite sub-optimal harvests and these demands, combined with the harvest failure of 1511, led to greater general indebtedness. Marginalized people, such as those living in the Bruche Valley (*das Breuschtal*), southeast of Strasbourg in the Vosges Mountains, began to ignore their fields in favour of living on wildlife and fish from the woodlands and waterways. In this specific example, the diocesan administration responded to peasant foraging by issuing a decree that ordered every horse-owning peasant to sow two fields, one of summer cereals and the other of winter cereals. Peasants with two horses would be obliged to cultivate double the area, and so on, while those without any horses were commanded to clear wood or scrubland to make them into arable fields. This decree effectively restored the volume of the tithe to the diocese by curtailing hunting and fishing as a means of subsistence, for the demands of sixteenth-century agriculture left little time for anything else.[13]

A cold winter throughout southwestern Germany in 1513 led into a cool summer with average harvests. In the context of weather-related poor or failed harvests for 1511, 1512, and 1513, heightened economic pressures from the region's elite such as those mentioned above resulted in widespread hardship among the peasantry and sparked the third *Bundschuh* rebellion in the autumn of 1513. Centred in the southern reaches of the Black Forest, conspirators led by Joss Fritz aimed to capture Freibourg-im-Breisgau on the feast of Saint Martin (*11/21 November*). Some of the plotters acted precipitously, though, and once it was discovered, the rebellion was suppressed by the same coalition of territorial authorities as had acted together in 1493 and 1502. Several conspirators were tortured and executed, although Fritz escaped capture by the authorities once again.[14]

The list of demands developed by the 1513 *Bundschuh* rebels shows an increasingly keen awareness of the precise remedies needed to address the difficulties facing Alsatian peasantry. Along with the goal of eradicating usury through interest-free loans, the rebels aimed to limit prebendaries to a single post with a cap of 20 pounds per year of income; useless or objectionable religious houses were to be dissolved, and the superfluous wealth of those remaining would be confiscated for the benefit of community coffers. This list also saw the first demand for the freedom to hunt and fish along with free access to the forests and waterways, as well as the first appearance of Scriptural justifications for the rebellion.[15]

Winter of 1514 is reported as having been extraordinarily cold throughout southern Germany, from Munich to the Rhine River. The Rhine froze from Saint Martin's Day 1513 (*11/21 November*) until late January in 1514 so thoroughly that there are reports of wagon teams driving across it; at Basel, tables and chairs were placed on the icebound river and a meal was served, followed by dancing. Alongside the opportunities for transportation and revelry, though, the frozen rivers also meant that water mills stood still. Those who depended on the mills to grind their grain could not, and 'the mountain dwellers had to pickle the wheat in water and eat it instead of bread.' Cold weather persisted into the spring, and the verdure only emerged after the first week of May, but the wine harvest was abundant, indicating that a necessary level of heat was eventually achieved for the grapes to mature.[16]

Although the climatic travails of 1514 did not contribute to a peasant uprising under the *Bundschuh* banner, there was no shortage of violent conflict in the region and nearby. Armed rebellion broke out 15 km south of Colmar that year when 1500 commoners from the town of Rouffach and the nearby countryside rose up together against the episcopal bailiff, the *magistrat*, and the town council. Known as the *Butzenkrieg*, it was triggered by particularly egregious behaviour on the part of the *magistrat*; the conflict was successfully mediated by representatives from Basel. Tensions also escalated to open violence that year in Ferrette, southeast of Basel in the Burgundian Gate, where the *landvogt* of Alsace, Wilhelm II von Rappoltstein, imposed his mediation to resolve the issue. Further away, the *Armer Konrad/Armer Kunz* ('Poor Conrad') rebellion in Württemberg protested against increasing taxes to fund the Duke's extravagant lifestyle and territorial ambitions. As Tom Scott indicates in *Town, Country, and Regions in Reformation Germany*, the absence of ideological perspectives from these insurgencies does not indicate the absence of an underlying dynamic reflected in their unfolding.[17]

Identifying this dynamic as a growing cultural schism between peasants and elites, historian Georges Bischoff is explicit in his assertion that peasant culture showed a heightened self-awareness and strength during the late fifteenth and early sixteenth centuries that was materially based on the deepening realization by Alsatian peasants and their urban counterparts that their interests did not align with those of the region's financial and religious leaders. Moreover, it was easy to look to the south, where the political and military successes of the Swiss Confederacy easily fired up dreams of winning greater independence from social superiors who held tight control over Alsace's economy. Also worth noting as contributing to the increasingly different views of peasants and lords are the many instances of negotiations conducted between the two groups, both successful and failed. Both contributed towards higher levels of cooperation and solidarity among peasant individuals and communities while identifying the elite as an Other with different goals and views.

Winter 1515 was again cold, although there are no mentions in available records of dancing on the Rhine. New Year's Day saw a heavy snowfall in Strasbourg, with a stroke of lightning which hit the cathedral and shook the whole city. Summer conditions were difficult: in Strasbourg and in Thann, northwest of Mulhouse, chroniclers relate that there was rain almost every day from Saint John's Day (*24 June/4 July*) to Saint Bartholomew's Day (*24 August/3 September*). Flooding and harvest failure ensued, leading to widespread famine in the Upper Rhine Valley. Further afield, excessive rains and floods heightened prices from Switzerland to Franconia.[18]

The next year, 1516, the opposite weather conditions were found: after a dry spring, not a single drop of rain fell from Saint John's Day (*24 June/4 July*) until Michaelmas (*29 September/9 October*). Strasbourg took advantage of the drought to enlarge its defensive trench from the Fisherman's Gate to Krutenau. Regionally, the Dominican chronicler at Thann wrote that – following a spring comet – there was such a severe drought that the

root vegetables, green vegetables, hay, and grain failed. A document from
Hagenau simply mentions that the summer drought ruined vegetables and
plants. Unsurprisingly, famine came to Alsace for a second year.[19]

Following a long, cold, and hungry winter, disaster struck in the spring,
when, on 17 April 1517 (*27 April*) temperatures plummeted below zero
throughout the Upper Rhine Valley during a spring drought where, from
Fastnacht until May, it only rained twice. The freezing cold severely damaged spring grain and grape vines on both the hills and the river plain at
vulnerable points in their growth. A rogation was quickly organized by
the *Rat* in Strasbourg, pleading with God for good weather and a successful harvest, but the plants did not recover. To compound the catastrophe, a hailstorm in late June (*26 June/6 July*) along the Rhine destroyed
most of what growth had occurred. Hungry rural people came to Strasbourg, whose coffers were generous, but smallpox broke out in the city that
year and the combination of famine and disease led to over 4000 deaths
within the walls. There was an outbreak of bubonic plague in Mulhouse,
although it is not known how many died there or throughout the rest of
Alace. Although wheat was imported from Burgundy and France, prices
in the region were very high and there were food shortages from Pentecost
(*31 May/10 June*) until Saint Martin's Day (*22/21 November*). One chronicler called it, simply, 'the bad year'.[20]

Enduring weather-related famine for the third consecutive year, peasants
and poor urban-dwellers did not fail to notice that unsympathetic landowners were exacerbating the crisis by insisting upon their usual financial tithes,
taxes, and dues. Loans were not forgiven, even to the starving, and many
peasants lost everything. Rumours of a widespread new *Bundschuh* uprising circulated during the summer of 1517 from at least mid-May, until, in
early September, a priest confessing one of the rebels reported the names of
individuals involved. The leaders were identified as Joss Fritz, whose organizational skills had only sharpened since 1513, and Veltin de Fribourg. Conspirators were identified in over 40 localities owing allegiance to a variety of
ecclesiastical and secular landowners on both sides of the Rhine, including
a cluster of villages around Strasbourg (Geispolsheim, La Wantzenau, Schil-
tigheim, Weyersheim, Geudertheim, Herrlisheim), in the countryside around
Hanau and Saverne, but particularly in the foothills of the Vosges between
Molsheim and Barr, and throughout the Black Forest, albeit concentrated in
the Kintzig Valley southeast of Strasbourg.

The uprising was planned to unfold in stages: first on 8 September with
the capture of *Zehnstädtebund*-member Rosheim as a primary target, timed
to coincide with the high volume of traffic on the roads for a fair day at
nearby Saverne (when the movement of a large number of men would be un-
noticeable). Taking control of Rosheim would be followed by the capture of
Saverne and Obernai, also members of the *Zehnstädtebund*. A big fire on the
crest of the Kniebis, about 50 km east of Strasbourg, on 26 September was to
rally those on that side of the Rhine; on 4 October, a band of rebels was to

assemble in the woods near Riedseltz with the intention of capturing the Free Imperial City of Lauterbourg and nearby Hagenau.[21]

While the far-reaching network and clear schedule of insurrectionists may be a testament to the organizing capacities of Joss Fritz, the rebels' goals speak to the nature of the discontent in Alsace. Specifically, Fritz and his fellow insurgents sought to abolish all taxes and rents and to dissolve local tribunals and councils, since these were often convened or led by landowners or their representatives. The revolutionaries planned to proclaim a general refusal of all civil authority except that of the emperor and the pope, and they planned to kill or expel members of the nobility and knights. They intended to kill all those who resisted. Once the beacon fires were lit, the rebels would send two men to the Swiss Confederate states to seek help and advice, hoping to gain enough assistance to ensure that the endeavour succeeded.

Arrests and torture elicited enough information for the aforementioned coalition of landowners to grasp the threat they faced should the peasantry unite against them, and their militias suppressed the last *Bundschuh* insurrection before it could begin. The landowners, though, have not been the only people to understand the political and socio-cultural unity of the peasantry and most urban-dwellers: Georges Bischoff's historical analysis highlights that the peasants and the greater part of the population of the towns shared perspectives, experiences, and aims that made them 'a single mass, a great mass, but not a passive mass.' Almost 90% of the Alsatian population, the peasants and non-noble urbanites, in Bischoff's view, had developed an awareness of their own political culture and goals; they were able to express their opinions, and they were fully aware of their strength. This last rested on the military capacity of infantrymen, who, since the victories of the Swiss decades earlier, were recognized for their potential as masters of the field. Their shared religious culture, as well, brought them together; most lived close to lower-ranked clerics whose preaching emphasized egalitarianism and fraternity with neighbours. Bischoff argues convincingly that the peasantry were not seeking a return to ancient custom with the *Bundschuh* rebellions, but forward-looking change towards greater equality and fewer overlords.[22]

Following the suppression of rebellion in November 1517, the winter of 1518 was long and cold, with a snap of extreme cold in February across southern Germany. A very dry spring combined with a week of severe cold in early June to create the conditions for a fourth year of famine and death across Alsace. On 14 July 1518 (*24 July*), Frau Troffea stepped out of her house into the streets of Strasbourg and swayed from foot to foot with rapid, rhythmic, and stiff motions for several hours. This behaviour was identified by peers as 'dancing.' After collapsing with exhaustion, she enjoyed only a few hours of sleep before resuming her dance. After she had suffered this for six days, earning bruises and lacerations on her bloody feet, the Council of XXI ordered her to be taken to a shrine to St. Vitus at Veitsburg, about 50 km away near Saverne. Frau Troffea's final fate is unknown, but her problem proved contagious: within days, over 30 people had taken to the streets to dance and within a month, up to 400 people were manically dancing. The best medical advice available at the time suggested that the fiery energy of dance had to be depleted from the bodies of the afflicted, so they were gathered into a few locations and spurred to continue dancing with music, refreshments, and non-afflicted dancers hired *pour encourager les autres*. Many perished before the last week of August (the first and second weeks of September by the Georgian calendar) by dancing themselves to death, at which point the XXI changed their approach and ordered all the dancers to be carted to the same St Vitus' shrine; there, they were given small crosses and red shoes, and a mass was performed. A few more people succumbed after this mass removal, but as the 1518 harvest began and appeared to be sufficient, the dancing plague disappeared from the area.[23]

After initial suggestions by medical practitioners inflamed the situation, Strasbourg's XXI demonstrably accepted a spiritual understanding of the problem. Historian of medicine John Waller, however, understands this outbreak as an example of a hysterical reaction to unrelenting and severe stress located in a specific cultural context: a catharsis of misery, suggestion, and belief. He describes the dancers as rendered vulnerable to unexpected and unsought altered states of consciousness or spontaneous trance through malnutrition and high levels of psychological distress, in a culture where it was believed that individual saints would curse humans with specific maladies as a sign of their wrath. Waller specifically identifies the harvest failures of 1516 and 1517 and the ensuing winter famines as primary causes for the outbreak in 1518, along with the stress caused by the spiritual consequence of defaulting on loans (excommunication) and widespread dread that personal corruption of those responsible for mediating with the spiritual world rendered the sacraments useless. It is also likely that the repression of any possibility for widespread social change which would ameliorate the economic pressures, such as that sought by the unsuccessful *Bundschuh* rebels several months earlier, deepened the distress felt by Frau Troffea and others in her situation.[24]

The harvest of 1518 did indeed prove successful (if mediocre) and so did that of 1519, despite a severe thunderstorm near Thann at midnight on Saints Peter and Paul's Day *(29 June/9 July)*. Prices regularized and Alsatians did their best to recover from the earlier calamities. It is impossible to know exactly when rumours or foreign printed material about Luther and his theological insights began to arrive in Strasbourg, but we know that reprints of his tracts in the Free Imperial City began in 1519. The social and cultural impacts of Luther's work will be explored in the next section, yet it is important to remember that these effects occurred alongside the weather events described here.[25]

The same year in which Luther's views found widespread interest in Strasbourg, the weather was cold and wet in Alsace, with sub-zero temperatures in the spring and autumn. The wine harvest of 1520 failed and at Advent, Strasbourg and several villages around the city were flooded for eight days. The situation reversed in 1521, however, when a hot and dry season led to a plentiful harvest of good wine and grain (although the vegetable crops did not do as well). An outstanding quality of wine was produced in 1522 after late spring frosts led into a hot summer, although there was only low volume. The good weather continued during the growing season of 1523, when the chronicler of the convent at Thann noted that, on the whole, it was a fruitful year in wine, cereal, fruit, and hay, as well as vegetables. For three days around Saint Martin's Day (*11/21 November*), however, a deep cold landed in the region and much of the vine stock froze to death both on the hillsides and in the valleys.[26]

Bitter weather continued in winter 1524, with heavy rains and hail on Three Kings Day (Epiphany, *6/16 January*) throughout the Upper Rhine Valley which, along with strong winds, brought widespread flooding and heavy damage to property and crops. The rain, wind, floods, and damage came again around Easter (*27 March/6 April*); whether in the winter or spring deluge, Clemens Zyegler almost drowned. Destructive frosts regularly visited the region until Pentecost (*15/25 May*), and hail in July further damaged the vines and crops. With almost a year of weather-related damage to the vines and fruiting trees, harvests throughout 1524 were poor, stimulating high prices and, thanks to the combination with ongoing expectations for the tithe and taxes, famine throughout southwestern Germany. Refugees streamed into Strasbourg, where Evangelicals had re-organized the welfare system (more on this next chapter). By mid-summer, there were grumblings of rebellion in Stühlingen and a treaty of mutual support was agreed between the peasants there and the reform-minded citizens of Waldshut, both to the east of Alsace. A heavy-handed response by authorities to the treaty stimulated similar allegiances; with time to spare during a cold winter, material demands boosted by preaching (explored in the next section) solidified a regional network of peasants into revolutionaries. In spring 1525, German Peasants' War began; it ended in defeat.[27]

This section provides the weather report for the 16 years between 1509 and 1524 – the middle period of this volume's 68-year weather report – and some of the social consequences of that weather. To summarize: during these years, eight harvests were reported to be above average (1510, 1514, and 1523) and average (1513, 1518, 1519, 1521, and 1522). During the same span of time, two harvests were reported to be poor (1509 and 1524) and six harvests failed entirely (1511, 1512, 1515, 1516, 1517, and 1524).

The poorer population in Alsace and Strasbourg endured dearth (scarce food supply) twice from 1509 to 1524 (in 1512 and 1520), thanks to bad weather. Weather-related harvest failures led to famine (severe food shortage) in Alsace four times during the same period (1515, 1516, 1517, and 1524).

The combination of stress from the natural environment and economic pressures from social superiors led to plans for peasant rebellions in 1513 and 1517 and contributed to the participation of Alsatian peasants in the German Peasants' War, planned the winter after the harvest failure and famine of 1524 and launched in spring 1525.

Peasant culture in Alsace

Luther's earlier publications appeared to emphasize social equality, particularly *An den christlichen Adel deutscher Nation* ('To the Christian Nobility of the German Nation,' published August 1520), *Von der babylonischen Gefangenschaft der Kirche* ('Babylonian Captivity of the Church,' published October 1520) and *Von der Freiheit eines Christenmenschen* ('On the Freedom of a Christian,' published November 1520); the first, in particular, was regularly read as a call for violent reform. However, by April 1525 his tractate *Ermahnung zum Frieden auf die zwölf Artikel der Bauerschaft in Schwaben* ('An Admonition to Peace on the Twelve Articles of the Peasantry in Swabia') and, in May, *Wider die Mordischen und Reubischen Rotten der Bawren* ('Against the Murderous, Thieving Hordes of Peasants') made clear not only that he did not support the establishment of social equality through revolution, he actively despised the widespread community who had understood his theological reforms to include social as well as spiritual reformations. Godliness, wrote Luther, lay in social stability, which – for him and for many authorities – meant maintaining the social status quo. It is nonetheless obvious from peasant sources that for most (if not all) who participated in the *Bundschuh* rebellions and the Peasants' War of 1525, the Gospel was a central means of conceptualizing and organizing their plan to defy late medieval social hierarchies and establish a Godly society based on peasant cultural norms.

These different views of Godliness and righteous behaviour contributed to the deaths of thousands of peasants, illustrate the cultural power of divergent yet deeply held interpretations of the central text in question – the Bible – and indicate the presence of competing views of the world: that is, different (although mutually comprehensible) cultures. That people who held wildly dissimilar connotations of Christian righteousness were each fervently committed to creating or maintaining their understanding of a Godly community is obvious, and such deep differences do not arise quickly; they speak to long-lasting cultural development in different directions. Perhaps rather than Miriam Usher Chrisman's distinction between lay culture and learned culture, which was based on her linguistic analysis of the pool of publications in Strasbourg, then, a more relevant distinction may be made here between elite culture and peasant culture. Such a distinction is based on the differences between those who, like Geiler, found righteousness in the existing hierarchically ordered community and those who sought it in an egalitarian type of community organization. This section explores the case for peasant culture in early sixteenth-century Alsace, one within which such representations of nature as articulated in the following section would be commonly accepted as valid, useful, and reliable.[28]

Building on Peter Blickle's insightful work, Georges Bischoff offers a nuanced exploration of this peasant culture, one which incorporates the peasants' political goals and religious views. While acknowledging the economic nature of the demands issued by the *Bundshuh* rebels, when assessing them in their social and cultural contexts Bischoff sets aside determinist arguments from both Marxist and non-Marxist historians. He argues instead that the rebels were not regressively seeking a return to ancient custom, but, rather, a progressive solution to problems which would establish a fundamental equality among all the members of society. He warns explicitly against mistaking a lack of agency by peasants for the absence of a global vision, and asserts that the rebels were not located at the margins of an elite-driven evolution of society and politics, but, rather, at the core of a cultural transformation which was overtaking northern European culture and which was resisted by those who benefited from the status quo. Specifically, in explaining the relationship between the emergence of Protestantism and the Peasants' War, Bischoff persuasively argues that Luther did not produce a rebellion by the peasants; the revolution already in progress produced Martin Luther.[29]

Bischoff's assertions are based on three arguments: first, that to view the Alsatian peasantry as a gathering of small communities clutching at their customs and confined within seigneurial boundaries is to ignore the much more powerful factor of cultural unity among the peasantry and the artisanal classes in the cities. The vast majority of the population were peasants, and they shared political opinions which only partly overlapped with those of their overlords – views which found expression in the demands accompanying the frequent uprisings throughout southern Germany.

Another powerful element of that unity was their religious culture, the core of Bischoff's second argument. This element of peasant culture developed through close proximity to the lower orders and priests who had the most to gain from opposing the Roman Babylon. Their visions for the future were egalitarian, fraternal, and Christian: proto-reform, in essence. Luther's theological seeds fell onto extraordinarily fertile ground because his was the single conceptual leap needed to step out of the late medieval Catholic paradigm. Bischoff's third argument is that the means to establish the peasant vision of an ideal society was found in the infantryman, who had gained a new respect on the battle field since the Swiss victories against Charles the Bold in 1476–1477. Infantrymen were soldiers without the financial means or social status to become knights: armed peasants, in other words. Technological innovations in armaments gave these foot soldiers realistic confidence in challenging mounted knights.

The grounds of Bischoff's insistence that peasant culture generated Luther and his religious insights are convincing, as the reformer's personal background is intimately associated with peasant culture. His grandfather was a farmer in or around Eisleben, a small town northeast of Leipzig in eastern Germany. Martin's father, Hans, may have married above himself when marrying Margareta, but she remained the daughter of peasants – although wealthier than farm-born Hans. Martin was born in 1483, and Hans moved the small family a few kilometres away to Mansfeld less than a year later. With Leipzig over 110 km away from the small town, at least a 15-hour walk on today's roads, and Magdeburg the same distance away towards the north, there is no hesitation in identifying Mansfeld as nestled deep in the peasant culture of rural Germany. Early schooling was provided by the *Bruder für Gemeinsamen Leben* in Magdeburg, emphasizing heartfelt piety and humble (peasant-like) devotion. Seventeen years old when he started university at Erfurt, Luther's formative years would have been lived among his (admittedly upwardly mobile) peasant family and friends.[30]

In asserting that peasants, loosely defined, established the cultural parameters within which Luther had his religious epiphany, Bischoff is arguing here directly against Peter Blickle's conclusion that the Lutheran movement produced the Peasants' War, but also against older explanations of the Peasants' War as a rebellion against the early modern state or as part of an 'early bourgeois revolution' in Germany. In lieu of these explanations, Bischoff proposes the slow development of a unified consciousness which took place over 30 years among the peasant class. Notwithstanding Luther's repudiation of social transformation as a central element in a Christian reform, and the agreement of the Strasbourg Evangelicals with his stance, the peasants and their sympathizers continued to develop their view of Scripture parallel to developments in Evangelical theology and doctrine. The peasant viewpoint was defined as radical largely due to their insistence that the Gospel advocated social organization in a manner which legitimized the views and aspirations of the poorest in society.[31]

The conceptual core of peasant political consciousness was found in the collective exercise of sovereignty – the community's ability to administer its own affairs, if you will – and the need for a contract between the governed and the governors. The peasant vision of collective self-management upheld by the authority of a single overlord is seen in their 1525 attempt to establish direct relationships between their self-governing communities and a single civic authority, the Holy Roman Emperor, thereby following the example of the Free Imperial Cities and the Swiss in bypassing intermediaries. Radical religious consciousness similarly attempted to bypass spiritual intermediaries between the Christian community and God, regularly asserting the priesthood of all believers and legitimizing with popular attention the views of men such as Strasbourg physician and lay preacher Karsthans (John Maurer), botanist and theologian Otto Brunfels, or Clemens Zyegler. By 1524, in contrast, Evangelical preachers in Strasbourg were looking to the *Rat* to justify and uphold their religious supremacy, and a year later, negotiating on behalf of the magistrates with peasant rebels; they had successfully integrated themselves into the existing social order. The military defeat of the peasants discredited their religious views, but did not dispel them; for example, James M. Stayer considers the roots of Anabaptism to be in the pre-Peasants' War religious radicals.[32]

Although Bischoff's analysis is limited to Alsatian peasant culture, Thomas Robisheaux makes a strong argument that this culture was common to peasants across western Germany, from the Elbe River through Saxony to Alsace

and the Upper Rhine Valley. Compared to peasants east of the Elbe, in eastern France, and northern Switzerland, the material prosperity of western
German peasants was based in soils that were generally more fertile; consequently, population was denser, towns more numerous, the nobility weaker,
and peasants more secure in their hold on the land, thanks to higher levels of
inheritable tenure. The Rhine River, in particular, provided rich lowland areas for agriculture and stimulated trade, commerce, and the growth of cities.
As Robisheaux writes, local populations interacted with social, economic,
and political structures to produce stable agrarian societies. In the context of
acknowledged regional variation, the characteristic community was a 'compact village,' with houses, barns, and other buildings (larger villages had a
parish church at the centre) clustered at the centre of individual fields, common lands, woods, and pastures. In the late fifteenth and early sixteenth
centuries, planting was coordinated and regulated in a three-field crop system, and survival depended on diversifying resources, which meant keeping
livestock, draught animals, and, where possible, fish ponds. These compact
villages structured everyday peasant life around shared public spaces, parish churches, cooperative work routines, common management of land and
resources, along with news exchange and gossip at the well, the tavern, the
mill, or the kitchen.[33]

Luther's theological challenges to Roman Catholic doctrine developed in a
confident and stable peasant culture where the ambition of a majority was to
dissolve layers of complex and exploitative political and economic authority.
Whether a single temporal overlord in the person of the Holy Roman Emperor or a single spiritual overlord in the divinity of Jesus Christ, both remove
a ranked hierarchy mediating between the common man and the source of
authority in favour of equality within the community which is maintained
by a single overarching and powerful leader. Such structural similarities are
not the focus of this research and other than noting the issue, it will not be
pursued here, even though the development of ideas such as the community
of all believers or *Omnia sunt Communia* within peasant culture is an issue
which deserves more research. Moreover, since Evangelical reformers quickly
moved away from a social revolution which would have compromised their
own rank within the hierarchy, the primary role of Luther's insights to the
peasant rebellion was to advance and support the radical religious views of
thinkers who were prepared to find within Scripture the views which would
bolster widespread social transformation.

Of relevance here is the representation of nature within this peasant culture, and it is considerably different from that expressed by Johannes Geiler.
The manner in which peasant rebels referred to streams, lakes, ponds, forests, woodlands, meadows, and fields in the *XII Artikels* while articulating
specific demands for access to the same agrees with a religious view of the
environment expressed in comments, anecdotes, and references in Zyegler's
Ein fast schon büchlin. That is, peasants demonstrate a representation of
nature that is essentially agricultural, including views that Godliness may

be found in practical, even dirty work and that direct interventions with the environment, such as farm work, are appropriate opportunities to receive the presence of Christ in one's heart. These representations will be explored in the next section.

Case study 2: peasant representations of the natural world

A maelstrom of religious perspectives swirled around Strasbourg and the Alsace from 1509 to 1525, particularly from 1519 to the onset of war, and although representing the natural world was not a primary focus of attention, nature served as a frame and a backdrop for the communication of ideas. This section provides an in-depth study of two sources: *Ein fast schon büchlin in welche(n) yederman findet ein hellen und claren verstandt von dem leib und blůt Christi*, a pamphlet by the religiously radical gardener of Strasbourg, Clemens Zyegler, and the text of the *XII Artikels*, the list of demands circulated by peasants throughout the region on the eve of the 1525 Peasants' War. Based on these two examples of individual and collective representations of nature, a reasonably confident analysis may be offered.[34]

Zyegler's aim in writing *Ein Fast schon büchlin* was to articulate a cogent view of the Eucharist; his view is that Christ's two natures (human and divine) were connected during his earthly life, and that without melting together, they formed a fruitful communion. Zyegler believed, however, that what was true of Christ can become true of any believer since the institution of the Last Supper, in that the true reception of the body of Christ does not depend on the words spoken by another, but solely upon the faith of the one who communes. While his thoughts on the Eucharist are of interest for future developments of radical and Evangelical dogma in Strasbourg, for our purposes it is important to notice the vivid presence of the natural world in the pamphlet. Like Geiler's *Die Emeis*, although with an entirely different theological perspective, Zyegler's exposition of a religious subject includes a view of his concept of nature which may be perceived through the references, anecdotes, and setting of his teaching.[35]

After asserting his identity as a genuine gardener (which he was), Zyegler writes that he found a peasant, one Peter Bauer, working in a field and that the conversation they enjoyed became the substance of his little book. The first 44 pages consist of Zyegler's exposition and justification of his understanding of the Eucharist; the following six pages are a dialogue, where Peter Bauer asks questions or searches for clarity about Zyegler's ideas. The explicitly agricultural setting in which the dialogue occurs may have been Zyegler's fictional attempt to legitimize his views within the peasant community or may be the written record of an actual event which took place somewhere in an Alsatian field.[36]

In the booklet, peasants and their occupation as tillers of the earth and tenders of beasts appear in several small allusions and twice as an extended reference elaborating a particular point. The Bible is portrayed as the field where hidden treasure lies, one that it brings great fruit to men on earth. Information about righteous behaviours is to be spread among the peasants without mention of other groups. While explaining the role of God in the Eucharistic process, Zyegler asserts that not a single grain of wheat would bring fruit unless God had ordained it, nor would the first grain itself. In responding to Peter Bauer's concern about the errors made by youth, Zyegler writes that he is unable to tear out unbelief from each man's heart with a dung-fork.[37]

Zyegler gives several reasons for his having undertaken to spread the Word of God, one of which he found in Deuteronomy 22:1. He begins by paraphrasing the passage, wherein God's law asked the people of Moses to return straying oxen, cows, or sheep to their brothers. He then points out to Peter Bauer that since a man is much more important than a cow, so much more zealous everyone should be in leading a man to Christ. Zyegler acknowledges that recognizing his brother or sister as astray also means that he is himself going astray, but asserts that he has now found the right 'field,' or place to find spiritual nourishment. It therefore is suitable for his soul's salvation to show it to his brother and to observe what happens to him as a result.[38]

Hints of the Great Chain of Being can be found in the passage, in the hierarchical distinction Zyegler makes between men and fish. Of greater impact on the paragraph, though, is his use of domestic animals as the source domain of a lively metaphor about communal self-regulation of Christians. Including himself in this community, as someone who has 'found the right field,' he elaborates the concept of an egalitarian community, a metaphorical 'herd' of Christians, where although temporary leaders may emerge as guides, all are ultimately equal. Also, Zyegler places knowledge of beast herding in direct juxtaposition with Scripture, using both to illustrate a desired attitude. With this passage, the farmer's practical experience is established as central to knowledge of how a genuinely Christian person should behave in the community.

In a passage about the distinction between the letter and the spirit of Christ, Zyegler discusses what may happen when the living spirit of Christ is accepted into one's heart. Particularly relevant to this research is his assertion that such revelation may take place anywhere: chopping wood, mucking out a stable, washing the dishes, sweeping the house, going to the field, heading out to a meadow, or looking after the cattle in the field. Whenever such thoughts would be found, however the person may be occupied, then that person partakes in the body and blood of Christ – even if there is no priest, altar, nor any outward sign of the Eucharist.[39]

According to Zyegler, performing domestic tasks or the chores of agriculture assumes a Godly hue when the heartfelt spirit of Christ is present. This is an introduction of holiness to an area of human endeavour hitherto the very definition of mundane, with its repeating round of daily and dirty chores. Once again, Zyegler's articulation of an intimate connection between a Godly peasant's activity and the natural world is notable; the very tasks which degrade the peasantry in Alsatian society, condemning them as crude, lowbred, or uncouth, here assume a distinct value as an appropriate setting for an expression of faith, in obedience to God. By locating a sacramental experience in this agricultural setting, Zyegler implicitly renders the farm into a sacred place and nods towards a view of farming as holy work.

Support for this interpretation of Zyegler's views of the natural world – which is to say, that God gave it to man for agricultural purposes – may be found not only in a single document, but in the many versions of the *XII Artikels* which spread through southern Germany, including Alsace, after their composition in March 1525 by rebels in Swabia. The list of 12 items can be understood as something of an 'economic-political platform' in which the demands included were those upon which the rebels could find agreement.[40]

With all religious or Biblical references removed, the skeleton of the *XII Artikels* composed in Memmingen is as follows:

1 that each community should have the power and authority to elect and appoint its own pastor;
2 that the large tithe on grain should be administered by the church warden and pay for the pastor's salary, serve as a reserve for the poor, and contribute to defence costs, while the smaller tithes should be eradicated;
3 that serfdom should be eradicated;
4 that without adequate documentary proof of private ownership of streams, lakes, or ponds, they should be administered by the community;
5 that without adequate documentary proof of private ownership of forests and woodlands they should be administered by the community;
6 that relief from labour for the lords must be attained;
7 that there should be no new demands for labour from the lords;
8 that there should be fairly established rents on leases;
9 that there should be consistency in matters of justice and law;
10 that without adequate documentary proof of private ownership of meadows and fields, that they should be restored to community administration;
11 that all death taxes should be abolished; and
12 that amendments to the demands should be argued on the basis of Scripture.[41]

Not every item was included by every group of peasants confronting their overlords. The articles, for example, which were presented before the *Rat* of Strasbourg by the Neubourg peasant band (who based them on those of the Altorf band) excluded nos. 7, 8, and 10, while those brought to the negotiating table by the rebels of the Sundgau had increased to 24 (demands regarding the waters, forests, and pastures remained and were supplemented with specific items such as the eradication of the tax for clearing communal lands, expulsion of Jews, and dissolution of the convents).[42]

To strip the articles down to their skeleton, however, is to remove the conceptual framework of the people who composed them, precisely that which is under investigation here. Using Articles 4, 5, 8, and 10 as his basis, David von Mayenburg offers a too-brief exploration of the convergence of practical legal reasoning and utopian theological thinking in the context of the early sixteenth-century legal debate about the agricultural order. Von Mayenburg perceives the claims as a legal text consisting of demands made in good faith by a group of people with basic faith in written law and the standard procedure for resolving agrarian conflict, as reflected in the standard legal terms found in the four articles. With respect to the Scriptural framework of the *Artikels*, Von Mayenburg notes that deriving an argument from divine law, particularly Genesis, was widespread common practice in legal texts until the early modern period. It was not, he says, only an invention of the peasants, pointing to the *Sachsenspiegel* for demonstration, as well as to the *Reformatio Sigismundi*, a German-language document from the 1439 Council of Basel concerned with efforts to reform the Holy Roman Empire. Von Mayenburg asserts that the legal goals of the peasants were the core conceptual drivers of the Articles. Further supporting Bischoff's emphasis on the progressive nature of peasant culture, Von Mayenburg concludes that the Bible was not exploited in the *XII Artikels* with fundamentalist intentions, but rather, moderated demands for changes in existing agricultural law, even while it contributed to furthering the goals of reform-minded lawyers.[43]

The economic and legal agenda of the *XII Artikels* are not in dispute here. However, any given text may serve a variety of purposes and querying the religious framework within which they were couched focuses attention on values and perceptions of the natural world through which the peasants justified their demands for access to natural resources and on the basis of which they sought a common understanding with social superiors. Article Four, for example, provides a clear articulation of the rebels' assumptions about the relationship of God and the natural environment, and the appropriate relationship of human beings to both. That is, the prevention of commoners from catching wild game, wildfowl, or fish from the running waters is interpreted as improper, unbrotherly, selfish, and contrary to God's Word. With a direct appeal to Scripture, the authors reminded their target audience of the position Christian men occupied with respect to the rest of creation, which is to say, dominant over all animals, birds, and fish. With this, an important medieval signifier of noble status – the inherent right to hunt, to hawk, to fowl, and to fish as desired – was effaced. As Christians, the rebels claimed, they had as much right to exercise power over animals as nobles (provided the nobility could not prove title).[44]

A specifically peasant perspective is found in the second sentence, which describes their plight as being obliged to suffer in silence while beasts gobble up the crops which God gave for man's use because, in some places, the rulers protect the game – a practice which offends both God and neighbour.

There are several points of interest about this comment; firstly, that the authors present crops as given by God for man's use. Agricultural goods require time, attention, and labour to produce, and yet, this view of them conceals the human effort involved to focus on the divine gift. By doing so, the authors' comment reminds the reader that farming is an activity which draws one into relationship with God, and, once again, nods to the notion that the successful farmer is doing the work God intended for a Christian man.

The pervasive power of this cultural conviction is revealed in a distant example: for hundreds of years, Europeans considered Indigenous people to be 'wasting' God's gift because, at the time of contact, they were not farming in a manner recognizable to the European explorers (the sustainability of Indigenous economies was invisible to Europeans). This so-called wastage was sufficient justification to declare many lands *Terra Nullius* and dispossess Indigenous people from their territories for centuries.

The view that God gave the Earth to man in order for agriculture to be introduced is supported by the second point of interest in the statement, which is that God is offended when the dumb beasts gobble up the crops. An offence is a transgression of the desired order, and for the peasants to consider the predations of wild animals on their crops as an offence to God is to consider the creation of that crop as divinely ordained; this implies, yet again, that the work that went into its creation is also desired by God. The third point of interest, then, identifies gobbling beasts as an offence against neighbours, whose Christian love of their brothers (and, implicitly, individual contributions to a communal fund for poor relief) would be tried if crops were destroyed. The demand in the Fourth Article for the peasants to hunt is not simply for legitimate access to game, but is framed in a manner which shows their belief that farming was divinely ordained work, that God desired them to succeed in it, and that its failure was a threat to the stability of a Godly community.

It should also be noted that the *XII Artikels* were introduced with passages establishing their justification through Scripture, and attention is drawn to the first demand, that each community should have the right to elect, appoint, or dismiss its own pastor. It is fair to suggest that the authors understood the economic aspects of each demand as the means to an essentially religious goal: the establishment of a Christian community on Earth, whose inhabitants would order themselves in an economically egalitarian manner before God. It was on behalf of the essentially religious righteousness of these demands that thousands of men risked (and lost) their lives.

The peasant view of nature which may be found in *Ein Fast schon büchlin* and the *XII Artikels* is agricultural. In essence, a Christianized natural environment was a farm or a garden. Religious virtue in the human interaction with nature is portrayed as being significantly determined by one's inner disposition, rather than by participation in a community ritual like the mass; therefore, activities such as farming are portrayed as righteous behaviour. When done with heartfelt devotion to Christ, farming and gardening assume sacred dimensions.

The Alsatian peasantry after 1525

The peasants' military defeat in 1525 had widespread social and cultural consequences. In the northern reaches of the Upper Rhine Valley, authorities co-operated across jurisdictions to restore the rural *status quo*, as shown by the programme adopted at Haguenau on 7 June 1525 (*17 June*) by the Holy Roman Emperor, the Bishop of Strasbourg, the Counts of Bitsch and Hanau, and the Free Imperial City of Strasbourg. A legal structure was established initially which intended to renew vows of obedience, confiscate the arms of any surviving peasant rebels, control religious worship, forbid the movement of serfs or supervise their movement while they repaired any damages, and to impose a general punitive tax, the *Brandschatzung*. An idea which had been discussed in June found fruition at a second meeting on 29 August (*8 September*) with the establishment of a common police force, to which each partner would contribute 20 knights and the possibility of adding another 50 at the request of the imperial prefect. Presided over by Phillip of Baden, a treaty signed on 18 September 1525 (*28 September*) formalized these agreements; its acceptance by rebels in Alsace avoided the extreme levels of violent suppression undertaken in Swabia or Thuringia on the orders of Archduke Ferdinand.[45]

As Bischoff notes, though, disarming peasants was not enough to break them of their disobedience to the desires of the nobility; their independent customs, culture, and identity must also be destroyed. Throughout Alsace, religiously radical views were banned and there was a mandatory return to orthodox Roman Catholic practices. Included among the edicts of 29 August 1525 (*8 September*), above, were strict prohibitions against certain forms of sociability, such as shooting clubs or certain games involving the election of a 'king,' and limits on others, such as the number of guests at weddings, *Fastnacht* visiting, or baptismal display. Sarcastic or ritual reversals, such as declaring a boy to be Bishop and seating him in the cathedral, were forbidden, no matter how long the tradition had endured; pipes and drums were silenced and dancing was banished. Roman Catholic priests were to be selected and paid by civil authorities and were expected to address their topics in such a way as to inspire subjects towards brotherly love, praise of God, and obedience to temporal authority. Mass was to be offered in Latin, and all the tithes, great and small, were to be regularly collected.[46]

Similar terms were imposed on southern Alsace and the Breisgau. For example, at Saint-Hippolyte, a village in the Vosges Mountains southwest of Sélestat, public submission and vows of obedience by the peasants were followed by the surrender of any remaining arms and commands were issued to observe each day of fasting and abstinence, to pay all tithes, and to notify authorities of any who neglected to perform their duties. This last contributed to two waves of judicial punishments which swept the southern portions of Upper Rhine Valley; prominent rebels who survived the War were executed through hanging, drowning, or other means, while less prominent rebels suffered mutilation. The centre of activities was Ensisheim, the Austrian regency's seat of power in the area; Archduke Ferdinand also led, albeit vicariously, a rigorous overhaul of the criminal code, in general rendering penalties more severe.[47]

Alongside the bodily consequences of rebellion, the financial costs were high. The Duke of Lorraine's army pillaged and stole during its visit; prisoners were taken (as many as a quarter of the inhabitants of Marmoutier, for example) with high ransoms demanded. Some of the money given as ransom payments stayed within the region through allocation to territorial landowners such as the Archbishop of Wissembourg. Other forms of financial compensation were demanded by the victors, such as the collective fine imposed upon the village of Dettweiler. Included among the possessions of Strasbourg, this village consisted of 60 hearths; authorities determined all the villagers had participated in the rebellion in some way. The only exceptions were those widows who had lost their husbands before the uprising – and only if they did not have a son among the rebels. Other authorities, such as William von Hohnstein, Bishop of Strasbourg, levied blanket fines on villages throughout the rebelling areas.[48]

In the southern end of the Upper Rhine Valley, where the Holy Roman Emperor was a significant landowner, punishing fines were also a central means of continuing to subjugate the population. The approach, however, was noticeably more uniform and regimented: after enumerating each house in the cities and villages which were under the direct authority of the house of Austria or their vassals, a six-florin fine per guilty person was exacted by an administrative body specially created for the purpose. The heirs of a dead rebel were equally responsible for paying the fine. Bischoff notes that this process, which was carried out also in the region of Württemberg and elsewhere, represented a true revolution in that it marked the 'uniformization' of territorial income: the same people performed the job against the same criteria, for the single purposes of enriching the prince. Formerly, imperial income from peasants had been mediated by the competence and integrity of regional vassals, but this committee had been established directly by Archduke Ferdinand. Information gained in the census was archived at Ensisheim or Innsbruck and could have been the basis of other demands, but was lost; only the 1527 records pertaining to the city of Masevaux, west of Mulhouse, continue to exist.[49]

Indebtedness in the region, widespread among the poor before the Peasants' War, became burdensome throughout peasant society afterwards. Agricultural activities were interrupted by the devastating losses of population or by the conscription of labour to repair damages, which made bringing in the hay or gathering grapes for wine difficult to achieve. Institutions which depended on revenues from the harvest also suffered, such as the Abbeys of Andlau or Neubourg. The survival strategies of these institutions were similar to those of the peasants, which included selling future tithes or parts of their land base while pleading for leniency from their creditors.[50]

Long-term consequences of the Peasants' War for the Upper Rhine Valley include a notable concentration of power in the hands of a single authority: Archduke Ferdinand, Regent of the Austrian hereditary lands of the Hapsburgs on behalf of his brother, Emperor Charles V. However, this was not the only result, and the peasants' defeat was not an unmitigated disaster. For example, the ambitions of lordship were arrested, as were also those of the authorities of local institutions. The elite, whether religious, civil, or noble, did not receive any new privileges; sovereign authority contented itself with a monopoly of power exercised through legislation and judicial arbitration. This last was translated into such expressions as the 1532 *Constitutio Criminalis Carolina*; known in German as the *Halsgerichtsordnung*, it unified the legal system of the Holy Roman Empire, which until then, had operated haphazardly across jurisdictions. Along with this, intermediate authorities were divested of their powers; serfdom began to erode and disappear. Bischoff notes that an administration began to grow which was less accessible and, therefore, less malleable – a direct ancestor of the modern state.[51]

What was true for most of the Upper Rhine Valley, though, was not true of the Free Imperial City of Strasbourg. Despite popular unrest in the city, the *Rat* did not rebel, support the rebels, or negotiate a compromise with them, but nor did it contribute to Duke Antoine's military victories. This passive stance served it well, as the city escaped direct chastisement from the Empire. Enduring tension within city walls, though, is evident through contradictory behaviour after the rebellion: as a landowner, the city participated equally with other landowners in repressing the population on its rural lands, but *Almosenschaffner* Lucas Hackfurt (see the next chapter for more information about this role) and Katharina Zell (wife of reformer Matthäus Zell) were assigned to organize shelter and food for some 3000 refugees from lordly retribution, including rebels, their wives and children, following the decisive battles of 1525. The exercise of authority inside city walls by the *Rat* was precarious, and as James Stayer notes, refugees in large imperial cities attracted less attention than persons who had stayed at home and embarrassed the same magistracy.[52]

One such embarrassment was Clemens Zyegler, who, although he did not participate in the military rebellion, was recognized as a pivotal preacher whose ideas inspired many to join the rebellion and who continued to influence radical thinkers such as Melchior Hoffman, Menno Simons, and Martin Steinbach. Zyegler himself seems to have become resigned to the social *status quo* after the peasants were defeated: he purchased citizenship in Strasbourg in December 1525, likely as an *Ausbürger*, and collected tithes for the poor and to support the Evangelical clergy. His writing shows a new orientation towards a visionary spiritualism focused on redemption through asceticism in this world and universal salvation in the next. For several years, Zyegler acted in a pastoral manner towards his co-parishioners, in competition with a *Rat*-appointed pastor; with the latter's death in 1528, the parish requested Zyegler for their pastor, but the *Rat* instead sent a part-time vicar.[53]

Religious unity was difficult to attain in Strasbourg following the Peasants' War; an indication of their over-riding focus on peaceful stability and their reluctance to embrace a new version of papist intolerance, the *Rat*'s decision to support Evangelical-style reform did not, for several years, go so far as to exclude religious radicals. As a result, Strasbourg became known as a haven for those who could accept neither Roman Catholicism nor Lutheran reforms, threatening Evangelical supremacy within the walls. The Evangelical community, led by Martin Bucer, enlisted the *Rat* for further support and an erratically enforced mandate was issued in July 1527 forbidding the populace from lodging or contacting Anabaptists, as many radicals had become known for their refusal of infant baptism (Zyegler, deepening a spiritualist and visionary orientation, was no longer among them). When the *Rat* abolished Mass from the city in 1929, Roman Catholics began to print and publish pamphlets, joining the flood of angry words brought to Strasbourg along with the refugees. Until 1533, alongside the Evangelical and Catholic communities, seven more-or-less distinct nonconformist groups were active in the city, aligning themselves along sectarian, spiritualist, and apocalyptic lines.[54]

In 1533, the Strasbourg Evangelical Church held a Synod to establish municipal doctrine, a disciplinary ordinance, and a system of visitation for rural churches. Bucer hoped that working out a confession of faith would stop the radical movements and implement a moral and ecclesiastical code for Strasbourg's urban and rural territories. Zyegler's influence was acknowledged when allocating him and his disciple, Martin Stör, first and second place in the examinations of selected nonconformists on 11 June 1533 (*21 June*). Under direct questioning by Bucer, Zyegler was accused of denying the existence of hell, Satan, and the Last Judgment, and – ironically – with teaching that God could be served through violence. Successfully defending himself, Zyegler was declared not a danger to the city and released; he was allowed to remain in Strasbourg providing that he did not preach.[55]

Frightened by the behaviour of Anabaptists in Münster, Strasbourg's *Rat* issued a series of statues against radicals in 1534: in March of that year, Evangelical doctrine was declared to be official municipal doctrine, and in April, anyone unwilling to accept it was given one week to take their family and leave. In June, the *Rat* condemned the views of radicals such as Clemens Zyegler and all nonconformists were ordered, upon oath, to shun separate meetings and to perform militia service – or to leave. Returnees would be punished, and all nonconformist civil servants would be dismissed and expelled if they retained their errors; children were to be baptized no later than six weeks after birth. With these measures, the Evangelical battle in Strasbourg against Roman Catholics and the religious radicals was won.[56]

Despite Zyegler's avowed pacifism and repeated pleas from his neighbours for him to be appointed as their pastor, he was marginalized by Strasbourg's *Rat* for the rest of his life and refused any post. Perhaps with reason, from the *Rat*'s perspective: although he conformed outwardly, Zyegler understood the anger in his community and his later visions spoke of violence in the city and the fall of the emperor. Clemens Zyegler, likely around 65 years old at the time, died in 1552.[57]

Conclusions

Comprising the vast majority of the population in early sixteenth-century Alsace, peasants – both rural and the artisanal urban class – were clearly developing a culture distinct from that of their social superiors. The *Bundschuh* rebellions and the 1525 German Peasants' War are some social expressions of that culture; if these had been successful, cities, towns, and eventually entire regions would have been organized according to peasant values and priorities. Luther's religious insights initially appeared to provide theological foundations for this cultural and social transformation, but the eagerness of the common man in Alsace and elsewhere in the German-speaking lands to re-order society was met with military opposition and defeat.

Despite their commitment to the social goals they held, then, radically religious peasant rebels failed to transform the hierarchy of political and economic relationships at the core of Alsatian society. Nevertheless, the enduring cultural influence of their views can be seen in at least three areas: the growth of Anabaptism, the further development and consolidation of the earliest theological developments of Protestantism in Strasbourg, and the introduction of new religious views of the natural world. The first two have been very well addressed by James M. Stayer and Amy Nelson Burnett; a discussion of the third, then, is opportune here.[58]

Exploring this situation from the perspective of the hybrid and interactive model of socio-economic metabolism, documentary evidence for weather having caused stress upon Alsatian society between 1509 and 1525 is plentiful. Along with harsh economic demands and social restrictions by the region's economic and political elite, weather played a significant role in causing the famines of 1511, 1512, 1513, 1515, 1516, and 1517, as well as provoking a hysterical stress response in Strasbourg in summer 1518. Hysteria was not the only social response by Alsatians to the natural stressors: after two instances of three consecutive years of famine (1511, 1512, and 1513, as well as 1515, 1516, and 1517), peasants prepared themselves to rebel against social systems which oppressed them, since repeated rogations, prayers, or cloud exorcisms by Roman Catholic priests were not bringing mercy from God with regards to the weather. In the autumns of 1513 and 1517, increasingly elaborate plans were laid to overturn the social order in order to prioritize the individual and communal needs of the peasantry. At this time, although the Church was failing to provide successful ritual protection or meaningful spiritual succour to suffering peasants, there is every indication that, for the most part, peasant culture was not aimed at overturning Roman Catholic cultural authority before Luther sent his 95 theses to Albert of Brandenburg, Bishop of Mainz, on 31 October 1517.

Many peasants' acceptance of Catholic perspectives changed when Luther's religious revolution spread throughout southwestern Germany. His claim that the Roman Catholic tradition had misinterpreted the Bible, God's message to humanity, was bolstered by punishing weather conditions which were increasing in frequency and intensity since 1511 and visibly contributing to famine, misery, and death. Remember: from the Catholic perspective, weather could be used by God to punish sinners (see Chapter 2) and the failure of both rituals and rebellions as remedies created such despair and hopelessness among the Strasbourgeois that there was an outbreak of the dancing plague. Reception of Luther's message was strongest among those who suffered the most: the peasants and urban poor. From the perspective of the model of socio-economic metabolism, the cultural sphere of causation delivered an interpretive model which explained not only social but also natural conditions and which promised relief. That is, the fervour and eagerness with which Luther's views were accepted and promulgated among peasants are due, in part, to the natural stresses exerted upon their society by adverse weather conditions characteristic of the Spörer Minimum.

The strongest indication that the Catholic representation of nature lost interpretative authority in early sixteenth-century peasant culture came with changes in religious programmes which attempted to exert an effect upon the natural environment. That is, we know that Strasbourgeois simply stopped participating in communal rituals pleading with God for good weather. Successful rogations were held for protection against plague, war, and bad weather on 23 July 1511 (*2 August*); for good weather and a successful harvest on 13 May 1517 (*23 May*); and on 21 April 1519 (*1 May*) for the upcoming Imperial election, peace and harmony in the realm, and the growth of field crops. By 1524, however, the annual rogation on Saint Mark's Day (*25 April/5 May*) proved a humiliating exercise for the remaining devout Catholics: only a few people participated, while, at the same time, Hedio gathered a crowd for a homily inside the cathedral. Driving the point home, a listener locked the main passage into the building, forcing the rogation to re-enter through a side door. This refusal to participate in ritual rogations and sabotage of the same, however, may be reasonably attributed to the Lutheran challenge to Catholicism as much as to radical religious views held by members of the peasantry and their urban peers.[59]

With another weather-related harvest failure and famine in Alsace in 1524, the widespread conviction that God would continue to punish Christians until they created a society based on the Gospel inspired many hitherto uninvolved peasants to come together in rebellion. Accordingly, adverse weather conditions characteristic of the Spörer Minimum should also be included among the immediate causes of the 1525 German Peasants' War.

Although Luther betrayed peasant culture in 1525, his successful challenge to Catholic interpretive hegemony created a brief opportunity for other peasants to develop and share their perspectives as well. Documents produced by radical theologians, such as Zyegler's *Ein Fast schon büchlin,* and the

XII Artikels produced by the Memmingen peasants and copied throughout Alsace point to a change in the religious representation of nature which was well underway by the third decade of the sixteenth century. For such views to have been considered unexceptional by members of the peasantry in such distant locations as the foothills of the Vosges mountains and Memmingen in Upper Swabia implies they were perspectives which had been comfortably assimilated into German-speaking peasant culture by the end of 1524.

Rendering the farm and farming as central to a Christian vision of the world fully validated the practices of agriculture and, inadvertently, the intentional and unintentional impacts these practices had on the rest of the natural environment. The representation of agriculture found in *Ein Fast schon büchlin* and the *XII Artikels* was, if not explicitly holy, certainly a portrait of the farmer and the farm as central to God's ordered Creation. Moreover, the successful completion of agricultural work was envisaged as central to the calm activity of a Godly community. Although the rebels failed to re-order society in alignment with their understanding of a Godly Christian community, it is unlikely that this view of farming – the defining occupation of peasants – as holy work, beloved by God, simply faded out of reformist views; as will be seen in Chapter 4, it did, in fact, remain present and influential.

For these peasant authors, God's order was limited to the human agricultural community; the invasion of wild animals into agricultural space disrupted the order which farming imposed upon the untamed world, while the invasion of wealthy landowners disrupted the Christian social order intended by God. Their views were innocent of the Aristotelian natural philosophy taught in universities, and, perhaps for that reason, they lack a philosophical foundation which incorporates wilderness and wild animals as anything other than opponents. The enthusiasm of their convictions was based almost exclusively on their newly shared reinterpretations of the Bible, which, as the acknowledged central text of their culture, gave authority to their understanding and their actions.

Notes

1 Eric Vanhaute, *Peasants in World History* (New York and Abingdon: Routledge, 2021), 4. For further discussion about peasants and peasantries, see Tom Scott, ed., *The Peasantries of Europe from the Fourteenth to the Eighteenth Centuries* (London and New York: Longman, 1998); Theodor Shanin, *The Awkward Class: Political Sociology of Peasantry in a Developing Society: Russia 1910:1925* (Oxford: Clarendon Press, 1972), esp. Appendix A; Eric R. Wolf, *Peasants*, Foundations of Modern Anthropology Series (Englewood Cliffs: Prentiss-Hall, 1966); Werzner Rösener, *Peasantries of Europe*, trans. Thomas M. Barker, The Making of Europe Series (Oxford and Cambridge: Blackwell, 1994); and Werner Rösener, *Peasants in the Middle Ages*, trans. Alexander Stutzer (Cambridge: Polity, 1996).

2 Tom Scott, *The Peasantries of Europe*, 4; United Nations, General Assembly, *United Nations Declaration on the Rights of Peasants and Other People Working in Rural Areas*: Resolution/Adopted by the Human Rights Council on 28 September 2018, available online at https://digitallibrary.un.org/record/1650694 [accessed 6 February 2023].

3 Scott, *Peasantries of Europe*, 4–5.
4 The register of leases of Saint-Nicolas-aux-Ones in the City of Strasbourg Archives, K.N. fol. 218a, cited in Rudolphe Peter, 'Clement Ziegler the Gardener: The Man and His Work,' trans. Cynthia Reimer and John Derkson, *The Mennonite Quarterly Review*, 69 (1995), 426; John Derksen, 'Hans Adam and Jörg Ziegler: Strasbourg's Radical Tailors,' *The Mennonite Quarterly Review*, 15 (1997), 32.
5 City of Strasbourg archives, Book of Citizenship, vol. 1, col. 568, cited in Peter, 426; George Huntston Williams, *The Radical Reformation* (Philadelphia: The Westminster Press, 1962), 245; Clemens Zyegler, *Ein fast schon büchlin in welche(n) yederman findet ein hellen und claren verstandt von dem leib und blůt Christi* (Strasbourg: Johanem Schwan (?), 1525); Zyegler published at least six printed pamphlets between 1524 and 1532, all but one before the end of 1525, and four handwritten pamphlets from 1532 to 1552.
6 Derksen, 39–40.
7 Ibid., 35–7.
8 Ibid., 37.
9 Peter, 439–42; Derksen, 40.
10 'The Twelve Articles of the Upper Swabian Peasants,' *The Radical Reformation*, trans. and ed. Michael G. Baylor, Cambridge Texts in the History of Political Thought (Cambridge: Cambridge University Press, 1991), 231n1; Georges Bischoff, *La guerre des Paysans: L'Alsace et la révolution du Bundschuh 1493–1525* (Strasbourg: La Nuée Bleue, 2010),164; see, in particular, Peter Blickle, *The Revolution of 1525: The German Peasants' War from a New Perspective*, trans. Thomas A. Brady, Jr. and H.C. Erik Midelfort (Baltimore and London: The Johns Hopkins University Press, 1985), Chaps. 2 and 4.
11 Survey for 1509 and 1510 on the database www.Tambora.org, <www.tambora.org> [accessed 14 December 2022]; P.F. Malachias Tschamser, ed., *Annales oder Jahrs-Geschicthen der Baarfüferen oder Minderen Brüdern S. Franc. ord. inßgemein Conventualen gevannt, zu Thann*, 2 vols (Colmar: Hoffmann, 1864), vol. 1, 717; Léon Dacheux, ed., *Fragments de Diverses Vieilles Chroniques*, Fragments des Anciennes Chroniques d'Alsace, IV (Strasbourg: Imprimerie Strasbourgeoise, 1901), 70; Frère Martin Stauffenberger, 'Les Annales des Frères Mineurs de Strasbourg, rédigées par le frère Martin Stauffenberger, économe du couvent (1507-1510),' ed. Rodolphe Reuss, *Bulletin de la Société pour la Conservation des Monuments Historiques*, IIᵉ Serie, 18 (Strasbourg: R. Schultz & Cie, 1897), 311; Tschamser, I, 721; Xavier Mossmann, ed., *Chronique des dominicains de Guebwiller* (Guebwiller: G. Brückert; Colmar: L. Reiffinger; Strasbourg: Schmidt et Grucker, 1844), 103–4; Stauffenberger, 313; Hans Stolz, *Die Hans Stolz'sche Gebweiler Chronik. Zeugenbericht über den Bauernkreig am Oberrhein*, ed. Wolfram Stolz (Freiburg: Edition Stolz, 1979), 123.
12 Survey for 1511 on the database www.Tambora.org, <www.tambora.org> [accessed 2 January 2023]; Dacheux, *Fragments IV*, 61; Mossmann, 104; Franz-Josef Mone, ed., *Quellensammlung der badischen Landesgeschichte*, 1 and 2 (Karlsruhe: C. Macklot, 1848), 1, 258; Tschamser, I, 725; Gabriela Signori, 'Ritual und Ereignis: Die Straßburger Bittgänge zue Zeit der Burgunderkriege (1474–1477),' *Historische Zeitschrift* 264 (April 1997), 318.
13 Survey for 1512 on the database www.Tambora.org, <www.tambora.org> [accessed 2 January 2023]; Francis Rapp, 'The Social and Economic Prehistory of the Peasant War in Lower Alsace,' trans. Bob Scribner and Gerhard Benecke, *The German Peasant War of 1525 – New Viewpoints*, eds. Bob Scribner and Gerhard Benecke (London: George Allen & Unwin, 1979), 53–4; Bischoff, 106.

14 Survey for 1513 on the database www.Tambora.org, <www.tambora.org> [accessed 3 January 2023]; Bischoff, 106–7.

15 Bischoff, 106–7.

16 Survey for 1514 on the database www.Tambora.org, <www.tambora.org> [accessed 4 January 2023]; Stolz, 125–8.

17 Tom Scott, *Town, Country, and Regions in Reformation Germany* (Leiden: Brill, 2005), 77–98; Bischoff, 106–7.

18 Survey for 1515 on the database www.Tambora.org, <www.tambora.org> [accessed 4 January 2023]; Daniel Specklin, *Les Collectanées de Daniel Specklin, chronique Strasbourgeoise du Seizième Siècle*, ed. Rodolphe Reuss, Fragments des Anciennes Chronique d'Alsace, II (Strasbourg: Librairie J. Noiriel, 1890), 486, Op. Cit 2208 (Pp. Sch); Mone, 1, 258; Mone, 2, 141; Johannes Stedel, *Die Strassburger Chronik des Johannes Stedel*, ed. Paul Fritsch (Strasbourg: Sebastian Brant-Verl, 1934), 113; Specklin, 489, Op. Cit. 2217.

19 Survey for 1516 on the database www.Tambora.org, <www.tambora.org> [accessed 4 January 2023]; Mone, 2, 141; Jean-Frédéric Hermann, *Notices historiques, statistiques et litteraires sur la ville de Strasbourg* (Strasbourg: F.G. Levrault, 1817), 250; Tschamser, I, 738; Bibliotheque Municipale de Haguenau, MS 5.2, 202.

20 Survey for 1517 on the database www.Tambora.org, <www.tambora.org> [accessed 6 January 2023]; *Fastnacht*, a period of carnival-type behaviour celebrated in Alsace from the Thursday before Ash Wednesday, occurred in 1517 from Thursday, 19 February (*1 March*), until Tuesday, 24 February (*6 March*); Mone, 2, 141; Stedel, 113; Bibliotheque Municipale de Haguenau, MS 5.2, 202; Luzian Pfleger, 'Die Stadt- und Rats-Gottesdienste im Strassburger Münster,' *Archiv für Elsässische Kirchengeschichte*, 12 (1937), 40; Signori, 318; Tschamser, II, 4; Dacheux, *Fragments* IV, 71, No. 3978. These deaths represent approx. 20% of the city's population, estimated to be just over 20,000 people, although that figure does not include refugees. Dacheux, *Fragments*, III, 202, No. 3237[d]; Stolz, 131; quoted in John Waller, *The Dancing Plague: The Strange, True Story of an Extraordinary Illness* (Naperville: Sourcebooks, 2009), 59.

21 Bischoff, 113–7.

22 Ibid., 113–4.

23 Survey for 1518 on the database www.Tambora.org, <www.tambora.org> [accessed 14 January 2023]; Waller, 75; Jean Wencker, 'Chronik und Zeitregister der Statt Strassburg,' in *Les Chroniques Strasbourgeoises de Jacques Trausch et de Jean Wencker*, ed. Léon Dacheux, Fragments des Anciennes Chroniques d'Alsace, III (Strasbourg: Imprimerie Strasbourgeoise and R. Schultz, 1892), 148, no. 3007.

24 Waller, 205, 59–67.

25 Tschamser, II, 8; Survey for 1519 on the database www.Tambora.org, <www.tambora.org> [accessed 14 January 2023]; Miriam Usher Chrisman, *Bibliography of Strasbourg Imprints, 1480–1599* (New Haven and London: Yale University Press, 1982), 285; Miriam Usher Chrisman, *Strasbourg and the Reform: A Study in the Process of Change* (New Haven and London: Yale University Press, 1967), 98, 301.

26 Surveys for 1520, 1521, 1522, and 1523 on the database www.Tambora.org, <www.tambora.org> [accessed 23 January 2023]; Johann Jacob Meyer, *La Chronique Strasbourgeoise*, ed. Rodolphe Reuss (Strasbourg: J. Noiriel, 1873), 123; Tschamser, II, 17; Mossman, 114 and 115; Tschamser, II, 27; Stolz, 135.

27 Survey for 1524 on the database www.Tambora.org, <www.tambora.org> [accessed 23 January 2023]; Tschamser, II, p. 28; Claude Muller, *Chronique de la Viticulture Alsacienne au XVIe Siècle* (Riquewihr: Éeditions J.-D. Reber, 2005), 62.

28 Miriam Usher Chrisman, *Lay Culture, Learned Culture: Books and Social Change in Strasbourg, 1480–1599* (New Haven and London: Yale University Press, 1982), 167–9.

29 Bischoff, Ch. 12; the next few paragraphs rely on his analysis; Ibid., 296, 113; Scott, *Town, Country, and Regions*, Ch. 3; Bischoff, 114–5.

30 This very brief biography of Martin Luther is indebted to Richard Marius, *Martin Luther: The Christian Between God and Death* (Cambridge: Harvard University Press, 1999), Ch. 2, Ulinka Rublack, *Reformation Europe* (Cambridge: Cambridge University Press, 2005), 12–3, and 'Martin Luther's Childhood and Youth,' <http://www.luther.de/en/geburt.html> [accessed 31 January 2023].

31 If one were to accept Chrisman's division of southwestern Germany into six classes (nobility, urban elite, learned civil servants and professionals, minor civil servants and technicians, common burghers and artisans, peasants), the Bundschuh movement is clearly based in the culture of the last, with growing representation from the others in descending order, excepting the nobility and urban elite (whose views are seen in laws and ordinances). Miriam Usher Chrisman, *Conflicting Visions of Reform: German Lay Propaganda Pamphlets, 1519–1530* (New Jersey: Humanities Press, 1996), 6, Günther Franz, *Der deutsche Bauernkrieg*, 12th edition (Darmstadt: Wissenschaftliche Buchgesellschaft, 1984) or Adolf Laube, Max Steinmetz, and Günter Vogler, *Illustrierte Feschichte der deutschen frühbürgerlichen Revolution* (Berlin: Dietz, 1974). Assuming the appearance of the *Bundschuh* rebellions to be associated with the onset of Alsatian peasant culture may be a false correlation (said culture is likely to be much older), but that is a question for further research.

32 Bischoff, 296; For examples of Maurer's views, see Manfred Krebs and Jean Rott, eds., *Quellen zur Geschichte der Täufer*, Vol. 7, Elsass, I: Stadt Strassburg 1522–32 (Gütersloh, 1959), Nos. 1–4; James M. Stayer, 'The Radicalization of the Social Gospel of the Reformation, 1524–1527,' in *The German Peasants' War and Anabaptist Community of Goods* (Montreal and Kingston: McGill-Queen's University Press, 1991), 45–60.

33 Thomas Robisheaux, 'The Peasantries of West Germany, 1300–1750,' in *The Peasantries of Europe from the Fourteenth to the Eighteenth Centuries* (Longman: London and New York, 1998), 111–4.

34 Similar work has been undertaken on behalf of post-Conquest English peasantry; see Susan Kilby, 'A Different World? Reconstructing the Peasant Environment in Medieval Elton,' *Medieval Settlement Research*, 25 (2010), 72–7.

35 Zyegler is known to have published at least six printed pamphlets between 1524 and 1532, all but one before the end of 1525, and four handwritten pamphlets from 1532 to 1552. Peter, 422–6; see also Martin Arnold, *Handwerker als theologische Schriftsteller: Studien zu Flugschriften der frühen Reformation (1523–1525)*, Göttinger theologische Arbeiten, XXXXII (Göttingen: Vandenhoeck & Ruprecht, 1990), 106–44; John Derksen, 'Non-Violent Political Action in Sixteenth-Century Strasbourg: The Ziegler Brothers,' in *The Mennonite Quarterly Review*, 78 (2004), 543–56.

36 Zyegler, 8.

37 Ibid., 52.

38 Ibid., 15. Deuteronomy 22:1 *non videbis bovem fratris tui aut ovem errantem et praeteribis sed reduces fratri tuo.*

39 Zyegler, 32.

40 For a comprehensive exposition of all the articles, See Görge K. Hasselhoff and David von Mayenburg, eds., *Die Zwölf Artikel von 1525 und das "Göttliche Recht" der Bauern – rechtshistorische und theologische Dimension*, Studien des Bonner Zentrums für Religion und Gesellschaft, 8 (Würzburg: Ergon, 2012), as well as Chapters 2 and 3 of Blickle, *The Revolution of 1525*; Bischoff, 165.

41 Summarized from Blickle's translation; Blickle, *The Revolution of 1525*, 195–200.
42 Bischoff, 165, 214–5.
43 David von Mayenburg, 'Bäuerliche Beschworden als Rechtstexte: Die Artikel der oberdeutschen Bauern zur Agrarrechtsordnung (Art. 4, 5, 8 und 10),' in *Die Zwölf Artikel von 1525 und das "Göttliche Recht' der Bauern-rechtshistorische und theologische Dimensione*n, eds., Görge K. Hasselhoff and David von Mayenburg (Würzburg: Ergon-Verlag, 2012), 99–130.
44 Memmingen City Archives, Series A, No. 2 materials for Memminger city history, Series A No. 2, 2000.
45 Bischoff, 406.
46 Ibid., 411–2.
47 Ibid., 412–3, 417–20, 424–5.
48 Ibid., 430, 433.
49 Archive Municipale Masevaux, series EE 4, cited by Bischoff, 437.
50 Bischoff, 438.
51 Ibid., 438 and 453.
52 Rudolphe W. Heinze, *Reform and Conflict,* Monarch History of the Church, IV (Oxford: Monarch Books, 2006), 293; James M. Stayer, *The German Peasants' War and Anabaptist Community of Goods* (Montreal and Kingston: McGill – Queen's University Press, 1991), 77.
53 John Derksen, *From Radicals to Survivors: Strasbourg's Religious Nonconformists over Two Generations, 1525–1570*, Bibliotheca Humanistica & Reformatorica 61 (Utrecht: Hes and de Graaf, 2002), 38–42.
54 Derksen, *From Radicals to Survivors*, 61.
55 Ibid., 77–81.
56 Ibid., 83–4.
57 Ibid., 203 and 258.
58 See Stayer's work and Amy Nelson Burnett, *Karlstadt and the Origins of the Eucharistic Controversy: A Study of the Circulation of Ideas*, Oxford Studies in Historical Theology (New York: OUP USA, 2011).
59 Pfleger, 'Die Stadt- und Rats-Gottesdienste, 39–40; Stedel, 93.

References

Primary sources

Baylor, Michael G., *The German Reformation and the Peasants' War: A Brief History with Documents*, The Bedford Series in History and Culture (Boston and New York: Bedford/St. Martin's, 2012).
Dacheux, Léon, ed., *Fragments de Diverses Vieilles Chroniques*, Fragments des Anciennes Chroniques d'Alsace, IV (Strasbourg: Imprimerie Strasbourgeoise, 1901).
Memmingen City Archives, Series A, No. 2 materials for Memminger city history, Series A No. 2, 2000.
Meyer, Johann Jacob, *La Chronique Strasbourgeoise*, ed. Rodolphe Reuss (Strasbourg: J. Noiriel, 1873).
Mone, Franz-Josef, ed., *Quellensammlung der badischen Landesgeschichte*, 1 and 2 (Karlsruhe: C. Macklot, 1848).
Mossmann, Xavier, ed., *Chronique des dominicains de Guebwiller* (Guebwiller: G. Brückert; Colmar: L. Reiffinger; Strasbourg: Schmidt et Grucker, 1844).
Specklin, Daniel, *Les Collectanées de Daniel Specklin, chronique Strasbourgeoise du Seizième Siècle*, ed. Rodolphe Reuss, Fragments des Anciennes Chronique d'Alsace, II (Strasbourg: Librairie J. Noiriel, 1890).

Stauffenberger, Frère Martin, 'Les Annales des Frères Mineurs de Strasbourg, rédigées par le frère Martin Stauffenberger, économe du couvent (1507–1510),' ed. Rodolphe Reuss, *Bulletin de la Société pour la Conservation des Monuments Historiques*, IIᵉ Serie, 18 (Strasbourg: R. Schultz & Cie, 1897).

Stedel, Johannes, *Die Strassburger Chronik des Johannes Stedel*, ed. Paul Fritsch (Strasbourg: Sebastian Brant-Verl, 1934).

Stolz, Hans, *Die Hans Stolz'sche Gebweiler Chronik. Zeugenbericht über den Bauernkreig am Oberrhein*, ed. Wolfram Stolz (Freiburg: Edition Stolz, 1979).

'The Twelve Articles of the Upper Swabian Peasants,' *The Radical Reformation*, ed. and trans., Michael G. Baylor, Cambridge Texts in the History of Political Thought (Cambridge: Cambridge University Press, 1991).

Tschamser, P.F. Malachias, ed., *Annales oder Jahrs-Geschicthen der Baarfüferen oder Minderen Brüdern S. Franc. ord. inßgemein Conventualen gevannt, zu Thann*, 2 vols (Colmar: Hoffmann, 1864).

Wencker, Jean, 'Chronik und Zeitregister der Statt Strassburg,' in *Les Chroniques Strasbourgeoises de Jacques Trausch et de Jean Wencker*, ed. Léon Dacheux, Fragments des Anciennes Chroniques d'Alsace, III (Strasbourg: Imprimerie Strasbourgeoise and R. Schultz, 1892).

Zyegler, Clemens, *Ein fast schon büchlin in welche(n) yederman findet ein hellen und claren verstandt von dem leib und blůt Christi* (Strasbourg: Johanem Schwan(?), 1525).

Secondary sources

Arnold, Martin, *Handwerker als theologische Schriftsteller: Studien zu Flugschriften der frühen Reformation (1523–1525)*, Göttinger theologische Arbeiten, XXXXII (Göttingen: Vandenhoeck & Ruprecht, 1990).

Bischoff, Georges, *La guerre des Paysans: L'Alsace et la révolution du Bundschuh 1493–1525* (Strasbourg: La Nuée Blueu, 2010).

Blickle, Peter, *The Revolution of 1525: The German Peasants' War from a New Perspective*, trans. Thomas A. Brady, Jr. and H.C. Erik Midelfort (Baltimore and London: The Johns Hopkins University Press, 1985).

Burnett, Amy Nelson, *Karlstadt and the Origins of the Eucharistic Controversy: A Study of the Circulation of Ideas*, Oxford Studies in Historical Theology (New York: Oxford University Press USA, 2011).

Chrisman, Miriam Usher, *Strasbourg and the Reform: A Study in the Process of Change* (New Haven and London: Yale University Press, 1967).

——— *Bibliography of Strasbourg Imprints, 1480–1599* (New Haven and London: Yale University Press, 1982).

——— *Lay Culture, Learned Culture: Books and Social Change in Strasbourg, 1480–1599* (New Haven and London: Yale University Press, 1982).

——— *Conflicting Visions of Reform: German Lay Propaganda Pamphlets, 1519–1530* (New Jersey: Humanities Press, 1996).

Derksen, John, 'Hans Adam and Jörg Ziegler: Strasbourg's Radical Tailors,' *The Mennonite Quarterly Review*, 15 (1997), 31–43.

——— *From Radicals to Survivors: Strasbourg's Religious Nonconformists over Two Generations, 1525–1570*, Bibliotheca Humanistica & Reformatorica 61 (Utrecht: Hes and de Graaf, 2002).

—— 'Non-Violent Political Action in Sixteenth-Century Strasbourg: The Ziegler Brothers,' in *The Mennonite Quarterly Review*, 78 (October 2004), 543–56.

Franz, Günther, *Der deutsche Bauernkrieg*, 12th edition (Darmstadt: Wissenschaftliche Buchgesellschaft, 1984).

Hasselhoff, Görge K., and David von Mayenburg, eds., *Die Zwölf Artikel von 1525 und das "Göttliche Recht" der Bauern – rechtshistorische und theologische Dimension*, Studien des Bonner Zentrums für Religion und Gesellschaft, 8 (Würzburg: Ergon, 2012).

Heinze, Rudolphe W., *Reform and Conflict*, Monarch History of the Church, IV (Oxford: Monarch Books, 2006).

Hermann, Jean-Frédéric, *Notices historiques, statistiques et litteraires sur la ville de Strasbourg* (Strasbourg: F.G. Levrault, 1817).

Kilby, Susan, 'A Different World? Reconstructing the Peasant Environment in Medieval Elton,' *Medieval Settlement Research*, 25 (2010), 72–77.

Krebs, Manfred, and Jean Rott, eds., *Quellen zur Geschichte der Täufer*, Vol VII, Elsass I: Stadt Strassburg 1522–1532 (Gütersloh: Gütersloher Verlagshaus Gerd Mohn, 1959).

Laube, Adolf, Max Steinmetz, and Günter Vogler, *Illustrierte Feschichte der deutschen frühbürgerlichen Revolution* (Berlin: Dietz, 1974).

Marius, Richard, *Martin Luther: The Christian Between God and Death* (Cambridge: Harvard University Press, 1999).

von Mayenburg, David, 'Bäuerliche Beschworden als Rechtstexte: Die Artikel der oberdeutschen Bauern zur Agrarrechtsordnung (Art. 4, 5, 8 und 10),' in *Die Zwölf Artikel von 1525 und das "Göttliche Recht" der Bauern-rechtshistorische und theologische Dimensione*n, eds., Görge K. Hasselhoff and David von Mayenburg (Würzburg: Ergon-Verlag, 2012).

Muller, Claude, *Chronique de la Viticulture Alsacienne au XVIe Siècle* (Riquewihr: Éeditions J.-D. Reber, 2005).

Peter, Rudolphe, 'Clement Ziegler the Gardener: The Man and His Work,' trans. Cynthia Reimer and John Derkson, *The Mennonite Quarterly Review*, 69 (1995), 421–51.

Pfleger, Luzian, 'Die Stadt- und Rats-Gottesdienste im Strassburger Münster,' *Archiv für Elsässische Kirchengeschichte*, 12 (1937), 1–55.

Rapp, Francis, 'The Social and Economic Prehistory of the Peasant War in Lower Alsace,' in *The German Peasant War of 1525 – New Viewpoints*, trans. and eds. Bob Scribner and Gerhard Benecke (London: George Allen & Unwin, 1979).

Robisheaux, Thomas, 'The Peasantries of West Germany, 1300–1750,' in *The Peasantries of Europe from the Fourteenth to the Eighteenth Centuries* (Longman: London and New York, 1998).

Rösener, Werner, *Peasantries of Europe*, trans. Thomas M. Barker, The Making of Europe Series (Oxford UK and Cambridge USA: Blackwell, 1994).

—— *Peasants in the Middle Ages*, trans. Alexander Stutzer (Cambridge: Polity, 1996).

Rublack, Ulinka, *Reformation Europe* (Cambridge: Cambridge University Press, 2005).

Scott, Tom, ed., *The Peasantries of Europe from the Fourteenth to the Eighteenth Centuries* (London and New York: Longman, 1998).

—— *Town, Country, and Regions in Reformation Germany* (Leiden: Brill, 2005).

Shanin, Theodor, *The Awkward Class: Political Sociology of Peasantry in a Developing Society: Russia 1910:1925* (Oxford: Clarendon Press, 1972).

Signori, Gabriela, 'Ritual und Ereignis: Die Straßburger Bittgänge zue Zeit der Burgunderkriege (1474–1477),' *Historische Zeitschrift*, 264 (1997), 281–328.

Stayer, James M., *The German Peasants' War and Anabaptist Community of Goods* (Montreal and Kingston: McGill-Queen's University Press, 1991).

Vanhaute, Eric, *Peasants in World History* (New York and Abingdon: Routledge, 2021).

Waller, John, *The Dancing Plague: The Strange, True Story of an Extraordinary Illness* (Naperville: Sourcebooks, 2009).

Williams, George Huntston, *The Radical Reformation* (Philadelphia: The Westminster Press, 1962).

Wolf, Eric R., *Peasants*, Foundations of Modern Anthropology Series (Englewood Cliffs: Prentiss-Hall, 1966).

Digital sources

'Martin Luther's Childhood and Youth,' <http://www.luther.de/en/geburt.html> [accessed 31 January 2023].

United Nations, General Assembly, *United Nations Declaration on the Rights of Peasants and Other People Working in Rural Areas*: Resolution/Adopted by the Human Rights Council on 28 September 2018, available online at https://digital-library.un.org/record/1650694 [accessed 6 February 2023].

4 The reform of nature in Strasbourg (Case Study 3)

Introduction

From the broadest perspective, early sixteenth-century Roman Catholics did their best to maintain the supremacy of their centuries-old tradition in the Free Imperial City, along with the social hierarchy upon which it thrived. Meanwhile, radical proto-Anabaptist peasants fought to overthrow that social hierarchy in order to establish Godliness throughout the land. Evangelical Reformers took a middle route: they sought to address social problems through the introduction of cultural change. Inspired by Martin Luther's approach, Strasbourg Evangelicals read the Bible and challenged traditions and practices which – in their opinion – were not based on Scripture. We cannot know which religious perspective (if any) might have become pre-eminent if the peasants had not moved into armed rebellion, but once their uprising had been suppressed in Alsace, there was very little possibility their views would become dominant in Strasbourg. Instead, the city became a haven for religious diversity for around 15 years, during which period several interpretations of the Christian message lived in the *Freie Reichsstadt*. Over time, largely through working with politicians to bring to fruition their social ideals, the Evangelicals succeeded in making their perspective the dominant one in the city and, eventually, its official religious stance. This chapter, then, will focus on Reform representations of the natural world.[1]

The chapter begins with a brief introduction to the social life of Luther's Gospel in Strasbourg from 1517 to 1525 (which developed alongside the peasants' organization leading to rebellion in 1525), and a biography of Wolfgang Capito. The next section includes my final weather report, detailing weather events and some of their social consequences from 1525 to 1540. Case Study 3 ensues, which is an exploration of Capito's representation of the natural world in *Hexemeron Dei opus*, followed by a contextualization of that representation within the broader Evangelical community including Philip Melanchthon's reconciliation of reform theology with natural philosophy. The case study is followed by an outline of Jean Calvin's time in Strasbourg, a discussion of the relative absence of the natural environment in his 1539 edition of *Institutio Christianae Religionis* (*Institutes of Christian*

DOI: 10.4324/9781003262411-5

Religion), a brief discussion of the influence exerted by Capito and Calvin's views of the natural environment, and conclusions.

There are several reasons for foregrounding Capito and Calvin in this chapter: Capito is recognized as one of the four leading Strasbourg reformers, along with Matthäus Zell, Caspar Hedio, and Martin Bucer. Although politically and perhaps personally overshadowed by Bucer among his peers and historically, Capito's theological work continues to stand prominently among the Rhine School of exegesis. His *Hexemeron Dei opus*, an exegesis of Genesis that was very likely used as a textbook by students in Johannes Sturm's *Schola Argentoratensis* (see below for further discussion), is the focus of Case Study 3. I am taking advantage of a relatively brief stay by Calvin in the Free Imperial City to include a few words about his work. He lived in Strasbourg for only three years – from 1538 to 1541 – and undoubtedly had a much greater impact on the cultural and social life of Geneva. However, many scholars consider his time in Strasbourg to have contributed significantly to his theological and intellectual development and, given that the city was his home while revising *Institutio Christianae Religionis* from a catechism to the work of systematic theology which became the core of his enduring theological legacy, a brief foray into Calvin's views of nature is a worthwhile inclusion.[2]

While we can't know the precise date when news of Luther's defiance of the Roman church arrived in Strasbourg and Alsace, four of his works (one theological tract and three sermon collections) are known to have been printed on Strasbourg presses in 1519 and in 1520 printing of his treatises began in earnest. Around that time we catch isolated glimpses of the earliest preachers formally sharing Luther's view of the Word of God: Master Peter Phillips von Rumersberg in 1520, at L'Église de Saint Pierre-le-Vieux, and Tilman von Lyn, reader at the Carmelite cloister in 1521 (both dismissed from their posts by the Bishop of Strasbourg for their efforts). The successful introduction of the Pure Gospel was only made in 1521 by Matthäus Zell, a popular preacher also born in Kaysersberg and prominently installed in the *Liebfrauenmünster* chapel of *Sankt Laurentius*. The people's eager response to the Pure Gospel generated a request for him to occupy Geiler's stone pulpit, but the chapter refused, inspiring parishioners to erect a portable wooden pulpit which was assembled in the cathedral before each of Zell's sermons, then disassembled and carried home for protection. Zell stood firm in his convictions and soon inspired the highly respected Catholic humanist Wolfgang Capito to join him in the Reform. Capito's conversion, that of the third most important clerical figure of the city, was accompanied by that of Peter Wickgram, Geiler's nephew and Zell's eventual replacement in the chapel, among many others.[3]

One of the first areas where Evangelical influence may be seen is in the overhaul of Strasbourg's social welfare system or, as it was known then, relief for the poor. Pre-Reform poor relief was complex and varied all over Europe: traditional Roman Catholic models of charity emphasized person-to-person

Almosenspenden (alms-giving) on the basis that loving God through loving one's neighbour was a central means of expressing religious devotion and piety, along with seeking peace and order in the city. Numerous canonical decretals and the writings of Church Fathers viewed the withholding of personal assistance as sinful. As noted by Norman J. Wilson, the result was a generous but unsystematic giving of alms which did not discourage begging and which distributed money not to those in greatest need but to those who were boldest. Defining poverty was difficult moving target, beggars could become ubiquitous and dangerous, community boundaries shifted regularly, and there were never enough alms to meet the need.[4]

As a central institution of the period, the Church was not blind to inadequacies in addressing the problems of poverty. For example, Johannes Geiler von Kaysersberg, the author featured in Case Study 1 in this volume, preached about the inadequate system for distributing alms in Strasbourg and in 1501 proposed concrete solutions to the *Rat*. (They were ignored.) City administrators, however, regularly showed concern about the overwhelming begging: in 1464, they had reduced the time foreigners were allowed to beg in the *freie Reichsstadt* to three days, followed by a mandatory three-month absence. In 1506, this was reduced even further to a limit of one day and one night, which basically allowed for begging only by citizens who were members of Strasbourg's *Bettlerbrüderschaft* (beggars' guild). Rather than framing almsgiving as a means of earning personal salvation, by 1520, Martin Luther was expressing the view that one's love of God leads to an obligation to serve the needs of one's neighbours, and this was echoed by those in sympathy with his views.[5]

In 1523, the situation of the impoverished in Strasbourg changed dramatically: the *Rat* outlawed all forms of begging and poor relief was removed from Roman Catholic hands at the same time that the convents and monasteries of the city were being dissolved. The *Rat* took control of collecting funds, including installing alms boxes in churches and administering the money, even while they established stricter eligibility requirements. Only those who could not work would receive relief, and recipients were obliged to indicate their status with badges similar to those worn earlier by beggars so that their moral behaviour could be monitored. Notably, the civil authorities did not hand over control of the new welfare system to the Reformers, but instead, created the post of *Almosenschaffner*, or alms steward, a salaried position reporting to the city magistrates.[6]

The first *Almosenschaffner* was Lucas Hackfurt, a former priest and native of Strasbourg who had studied theology at Heidelberg. While directing a private Latin school in his home city, he was one of the earliest Catholic priests to accept Luther's teaching and became publicly vocal about the errors of the Roman Catholic Church. His meticulous weekly records show his developing views of the problems which came before him and provide a description of the social and economic conditions of the city at the time. He held his appointment as *Almosenschaffner* from August 1523 until his death in 1554;

his abilities were put to the test early, as harvest failure in 1524 led to famine in the region and a stream of refugees into the city. A further discussion of changes to the city's poor relief system is included in the weather report for 1529 and 1530 (below).[7]

By the spring of 1524, five of the nine parishes in Strasbourg had chosen to take Evangelical preachers in direct opposition to the will of church officials: Sainte Aurélie, Saint-Pierre-le-Jeune, Saint Stephan, Saint Martin, and Saint Laurent (to use their current names). The people who lived there were mostly artisans, labourers, gardeners, or other poor folk – the urban peasants, in effect. To evade legal consequences, as appointing preachers was the right of the Catholic Church (either bishop or chapter), parishioners from these parishes petitioned the *Rat* to administer their benefices. This request implied an unprecedented transfer of power to civil authorities and, in an exceptional turn of their decision-making process, the *Rat* consulted the full *Schöffen*, the 300 powerful guild leaders who sent representatives to Council, for a vote on the petition. On 24 August 1524, the latter voted that the *Magistrat* of the city should indeed be responsible for the appointment of parish priests; this was gratifying to the wide foundation of support for Reform-minded preachers in the city. Further developments of the Evangelical Reform movement in the *freie Reichsstadt* have served as a worthy subject for many scholars and interested readers are encouraged to visit sources listed in footnote 1.[8]

Before the 1525 peasants' rebellion, then, at least five interpretations of Christian doctrine vied for popularity in Strasbourg: conservative (Catholic), humanist, evangelical, radical (proto-Anabaptist), and spiritualist. Detailing their several overlapping perspectives about ritual behaviour, pedagogy, and pastoral care could, as Aubray notes, only contribute to confusion. In essence, they were distinguished by divergent views about the appropriate relationship of clergy to laity, the standards and enforcement of morality, and the definition and policing of orthodoxy. The period from 1525 to 1540, however, was one during which unity among the Evangelical community was slowly consolidating around the views of Martin Luther and Philip Melanchthon, championed in Strasbourg by Martin Bucer and Wolfgang Capito.[9]

Born at Haguenau in 1478 to Hans and Agnes Köpfel, Wolfgang Fabricius Capito attended the Latin school at Pforzheim, some 77 km away from the Free Imperial City. He then pursued his studies at the University of Freiburg-im-Breisgau, where he earned a Bachelor of Arts in 1505, a Master of Arts in 1506, and (after taking minor orders and being ordained as a priest in 1509) a Doctorate of Theology in 1515. It is not known precisely when he Latinized his name to Capito, but it is thought to have been at some point between 1506 and 1509, when he returned to Haguenau to work as a proof-reader.

As a Catholic intellectual, Capito is recognized as having had a nominalist outlook – that is, of following the *via moderna* along with Scotus, Occam, and Johannes Geiler von Kayserberg (whose sermons he very likely had occasion to hear). In 1512, having been appointed canon and preacher for the Benedictine foundation in Untergrombach (now Bruchsal), Capito began to learn Hebrew from a Jewish Spanish *converso* refugee; he integrated it and other ancient languages into his theological exercises, arguing that these alone allowed the scholar to return to the ancient sources. In 1515, Capito was invited by the Bishop of Basel, Christoph von Utenheim, to occupy the position of cathedral preacher there.

Once in Basel, Capito is known to have enjoyed the company of such noted humanists as Johannes Oecolampadius, Beatus Rhenanus, Konrad Pellikan, and, on occasion, Erasmus of Rotterdam. The combination of learning and piety shown by this last was a particularly powerful inspiration; Erasmus' *philosophia Christi* is described by Kittelson as having provided a complete intellectual and religious programme for Capito. A personal friendship also developed between the two while Capito assisted Erasmus with the Hebrew names for the *Novum Instrumentum*, and correspondence between them flourished for several years. Erasmus' public praise for Capito's skills in Greek, Latin, and Hebrew contributed to the latter's growing European reputation as a Hebraist.

For five years, Capito preached, taught, and lived the life of the mind in Basel, but in 1520 he accepted an invitation from Albert III von Brandenburg, Archbishop and Elector of Mainz, to become the cathedral preacher of the city. The appeal of Albert's invitation is obvious: as an archbishop and a ruling prince, the Elector of Mainz was the president of the Holy Roman Empire's Electoral College, arch-chancellor of the empire, and Primate of Germany. Although he had been appointed to preach in Albert's place, Capito was quickly relieved of those duties in favour of offering regular advice in theological matters to the most powerful German cleric of the time. This type of advice, in the years immediately following Martin Luther's 1517 challenge to the Roman Catholic Church, was a delicate matter. Capito had privately expressed doubts about the orthodox understanding of the Eucharist while in Bruchsal, in a conversation later reported by Konrad Pellikan, but at the time of his arrival in Mainz, he was still willing to stand with the Catholic community and work towards reform of the Universal Church. He was nonetheless in direct communication with Luther, who congratulated him warmly upon his appointment.

Luther had reason to be pleased with Capito's position at the elbow of the Archbishop of Mainz: within a few months of arriving in Mainz, Albert entrusted Capito with his response to a papal demand that the *Exsurge Domine*, the bull which summoned Luther to Rome, be published throughout Mainz and Magdeburg. The ensuing letter communicated agreement but delay, as Capito claimed (on Albert's behalf) that a consultation with the secular princes was necessary due to the support Luther enjoyed among the people. As a result, the *Exsurge Domine* was effectively killed as a means of blocking the spread of Luther's ideas. Capito took further advantage of his position to moderate and delay Albert's responses to several provocative publications, including Luther's 1520 pamphlet *Address to the Christian Nobility*. In 1521, Capito convinced Albert to refuse the post of Inquisitor

General for all Germany, thereby avoiding the introduction to the country of an equivalent of the Spanish, Portuguese, and Roman Inquisitions. Capito also threw dust in the eyes of Hieronymus Aleander, papal nuncio and leader of the opposition to Luther at the Diet of Worms in May 1521, by convincingly representing himself as loyal to Rome even while undermining the Edict of Worms in the same way he had scuttled the *Exsurge Domine*.

Capito's theological views from 1520 and 1521 are difficult to discern; Kittelson describes him as interested in Luther's ideas, but still an Erasmian humanist and, above all, dedicated to a peaceful and harmonious reform of the church. It was this spirit which led Capito to criticize Luther, both in letters and in person, even while protecting him from ecclesiastical officials. Letters to Erasmus reflect Capito's personal confusion. By February 1523, he was ready to leave Albert's service and move to Strasbourg, where his efforts to secure the provostship of the collegiate church of St. Thomas had been successful. He was granted citizenship by the *Rat* that same year.

Strasbourg in 1523, however, proved to hold little peace for Capito, as Reform controversies were alive and flourishing. A conversation with Matthäus Zell proved more influential than all of Luther's tracts and arguments, as, following the encounter, Capito began to express reformist views in his letters, tracts, and behaviour. On 2 March 1524, the parishioners of *Jung Sankt Peter*'s, one of Strasbourg's nine parish churches within city walls, invited Capito to become their preacher, and by 4 May, with the approval of the *Rat*, he was preaching there. It was a deliberate and public decision to break with the Hildebrandine tradition of hierarchical authority in favour of the Reformation principle of called ministry. His break with the Roman church was even more obvious to the Strasbourg community on 1 August, when he married Agnes Roettel, daughter of a member of the Strasbourg XV. By October 1524, Capito was referring to the Pope as the Antichrist; around the same time, he broke his friendship with Erasmus by publicly criticizing an early draft of *De libero arbitrio diatribe sive collatio* (The Freedom of the Will), the Dutch humanist's refutation of Martin Luther and his reforms.

Having made his commitment to Luther's reform, Capito brought his scholarly skills, personal tact, and political adroitness to the Evangelical community in Strasbourg. He authored numerous pamphlets (both anonymous and signed), theological tracts, and even hymns, as well as continuing to preach for several years and maintaining a large correspondence. After the Peasants' War, he joined Bucer, Hedio, and Zell in lobbying the *Rat* for a ban on the mass and other Catholic rituals, including rogations; their goal was achieved in February 1529. In the irenic spirit of the Gospel, he interviewed or extended personal hospitality to Anabaptists such as Michael Sattler, Jacob Kautz, and Martin Cellarius, among others, with the goal of assessing their relationship with the Holy Spirit and, if possible, converting them to Evangelical views. This led to some conflict with Martin Bucer, who was convinced that assessing the religiously radical on a case-by-case basis was dangerous to the Evangelical church in Strasbourg. Bucer preferred to have Evangelical orthodoxy defined in the city through a municipally sponsored synod featuring debates over the Sixteen Articles he had personally composed, followed by public vows of loyalty to the city and its religious doctrine; taken together, these measures would establish heterodoxy as a municipal crime. Capito's view of that approach, shared by other Evangelicals like Anton Engelbrecht, the pastor at *Stephanskapelle*, was that an individual conscience was not receptive to force in matters of faith. This latter perspective held sway for several years, largely due to the respect held for Capito by the *Rat*, and was the basis of Strasbourg's reputation as a haven for the religiously unorthodox during the period immediately following the Peasants' War.

Differences of opinion between Bucer and Capito regarding the most appropriate method of bringing order to the city's Christian community continued until 1531, when several grievous events brought about a change in Capito's outlook. These included the deaths of ally Huldrych Zwingli at the Battle of Keppel and close friend Johannes Oecolampadius in Basel. When Capito himself was infected by the plague in November and Agnes, his wife, was among the 1200 Strasbourgeois who died of the disease that autumn, he fell into depression. After his physical recovery, Bucer suggested Capito's spirits would find inspiration by accompanying Alexander Berner and other leading Strasbourg reformers on an extended multi-city tour of nearby reformed municipalities. The excursion was intended to find and foster organizational similarities among the new Evangelical churches, with a particular focus on their welfare systems. Capito's even-handed and skill in mediating conflict among the reformers of those cities were noted with appreciation. On the personal level, the demand and need for structural clarity appear to have impressed themselves upon him, as he returned to Strasbourg in early 1532 supportive of Bucer's approach to the issue of religious conformity. He also returned with a new wife, having married Wilbrandis Rosenblatt, the widow of Oecolampadius, in Basel (she would go on to marry Bucer after Capito's death). Bucer's long-anticipated synod took place in June 1533 with Capito's participation and full support.[10]

Having achieved primacy in municipal religious institutions, Strasbourg's Evangelicals then turned their attention to a central platform for establishing their vision of a Godly community: education reform. The first petition of 1524 from Evangelicals to the *Rat* (likely composed by Capito) had included an appeal for schoolmasters as necessary for the growth of piety. Theological education was a regular theme for Capito, as for many other reformers. It is seen in other petitions associated with him and was included in requests for access to income from secularized ecclesiastical properties, in the 1525 preface to the third edition of his 1516 Hebrew grammar (*Institutiuncula in Hebream linguam*), and in his willingness to advise the theological faculty at the University of Basel. He became deeply involved with the work of establishing a Latin school in Strasbourg, which found fruition in 1438 with the founding of the *Schola Argentoratensis*, directed by Johannes Sturm.

Students enjoyed Capito as one of the earliest lecturers and his exegesis of Genesis, *Hexemeron Dei Opus*, was published in 1539. The scholar became ill early in 1541, briefly recovered, but died of the bubonic plague on 2 November of that same year.[11]

Weather report from 1525 to 1541 and some social consequences of that weather

Despite flooding at the beginning of the Lenten fast, Alsatians in 1525 experienced a year with unexceptional weather conditions, but the extremely high social turbulence of the spring season led to a flood of refugees coming to Strasbourg. Katharina Schütz Zell, the preacher's wife, played a significant role in setting up a separate fund for refugees because they were not covered by the city's new welfare system; she became a key collaborator with *Almosenschaffner* Hackfurt in coping with the refugees. According to a chronicler at the Thann convent, the summer was very hot and dry, with varying (although sufficient) results for the harvest: garden vegetables failed, oats and barley were poor, but domestic fruit, wild fruit, and grapes for wine flourished. He makes particular note that the weather was so calm that the weather bells only rang twice to warn of storms. At nearby Gebweiler, Hans Stolz confirmed the Thann chronicler's comments, adding that although the oats and barley crops failed, other grains produced in plenty. Stolz refers to the area around Murbach as 'a virtuous land'.[12]

Following an extended period of heavy rain in late February 1526 (*early March*), high waters in the Vosges and Black Forest flowed down to the plain and flooded Strasbourg (although less severely than in 1480), provoking fears of a second Biblical Flood. April turned dry and the grapevines leafed out early and abundantly, although the dry heat arrived too quickly for farmers to perform the necessary task of opening up the soil around the vines. Despite this, harvests were good, with wine and grain so abundant that prices were low. Even bringing in the hay was supported with good weather. Weather in Alsace was equally unexceptional in 1527 and, despite a frightening comet on 11 October (*21 October*) and an early frost on Saint Luke's Day (*18/28 October*), harvests were abundant and of good quality. This was also the case for 1528, despite a rainstorm on 29 June (*9 July*), a hailstorm on 19 July (*29 July*), an epidemic in Strasbourg, and earthquakes in July and December.[13]

These four years of generally favourable weather conditions came to an end in 1529. On 18 April (*28 April*), so much snow fell around Freibourg that merchants had great difficulty leaving for markets in other cities. On 13 June (*23 June*), four days of rain led to severe flooding across the Rhine plain. Refugees came streaming into Strasbourg, where the able-bodied among them were put to work on the city's fortifications and the less fortunate found support from the *Almosenschaffner's* special fund. Along with water levels climbing to ten feet above usual in some places, landslides occurred at Rangen and Enchenberg in Lower Alsace, Guebwiller, Soultz, and Uffholz in

Upper Alsace (the last also losing its bridge), and Stauffen in the Breisgau. A Catholic chronicler at Thann commented that God had punished Basel, presumably for the city's recent acceptance of Protestantism, by making the waters so high that the workshops of goldsmiths and other artisans were flooded. More than a thousand sheep, cows, and pigs perished at Meyenheim and Merxheim (between Mulhouse and Colmar), and four people drowned at Bitschwiller, near Thann. The beasts of Colmar could not be brought to pasture for two weeks. The summer continued cold and rainy, which did not allow the grapes to mature and resulted in sour wine. Grains were moderately successful, but prices were high. Severe food shortages (famine) in the region drove more people to Strasbourg looking for succour.[14]

The winter following this cold, miserable summer was so temperate, however, that fieldwork could be done in one's shirtsleeves; there was no snow or ice during January or February 1530, and the trees began to bloom weeks earlier than usual. It was so warm, day and night, that it was reported to be more like summer than winter. However, on 4 April (*14 April*), the seasons caught up with a vengeance and snow fell, freezing the grapevines. Although harvests in the Upper Rhine were adequate, conditions of dearth again prevailed in Alsace after harvest failure in Switzerland led the Swiss to buy up grain supplies and the same conditions in Lorraine brought Lorrainians to Strasbourg and the Upper Rhine Valley as refugees.[15]

In response to these demanding years, *Almosenschaffner* Lucas Hackford's views about the degree of responsibility held by Strasbourg to the poor demonstrably changed. While in 1527 he had argued that all who appealed to the city should receive succour (foreigners as well as resident poor), in 1530 – after the dearth of the previous year – he recommended that foreigners should not be permitted to beg in the city. The *Rat* immediately accepted his advice and by September 1530 had issued the appropriate ordinances.[16]

Winter 1530 and summer 1531 followed the same pattern as the previous year: an unusually warm winter followed by freezing temperatures and snow from 9 April (*19 April*) to 14 April (*24 April*) 1531. After this difficult start, crops endured a hot and dry summer where regional waterways dwindled and some dried up entirely; flying insects were noticeably fewer. The evening of 4 September (*14 September*) saw a severe hailstorm break windows and mangle vines, trees, and gardens in southern Alsace from Guebwiller to Basel and the Breisgau. This difficult summer, bookended by destructive extreme weather events, combined with very high prices to create famine in Alsace in 1531, and the hospices of Strasbourg were crowded.[17]

In this year of famine, Hackfurt further refined his notion of the deserving poor (although this term was not used by him) by developing policy about the *Schultheisbürger*, those quasi-citizens who received alms but no political rights. Framing them as agitators, he recommended that if a *Schultheisbürger* was found begging in the street more than three times, their oath to the city should be revoked on the way to banishment from Strasbourg. The funds of the common chest, he contended, belonged to full citizens of the city, such as people incapacitated by illness, misfortunate youth who needed dowries or apprenticeship fees, or villagers who dwelt in direct possessions of the city. Foreigners who purchased access to the city's alms through becoming a *Schultheisbürger* were diverting money away from those to whom it properly belonged. Meanwhile, Strasbourg reformer Alexander Berner spearheaded a touring group (including Wolfgang Capito) to visit neighbouring cities in order to compare the capacities of their welfare systems under duress, visiting Nuremberg, Augsburg, Memmingen, Ulm, Isny, Lindau, St. Gall, Constance, Basel, Zurich, Baden, and smaller cities and towns throughout Württemberg. This implies that demand for charitable support – perhaps also related to challenging weather conditions – was widespread across southwestern Germany and northern Switzerland.[18]

The string of warm winters ended in 1532 with an abundance of snow which fell and lay deeply on the ground until 23 February (*4 March*), when warming temperatures brought thawing and melting. Trees and vines greened. However, on 15 April (*25 April*), a cold snap arrived and on 17 April (*27 April*), a hard frost and snow with a stiff wind followed by a cold, clear day froze all the plants. Despite this, harvests of 1532 were good, with an abundance of grapes, apples, pears, and grain throughout Alsace.[19]

Saint Valentine's Day in 1533 did not usher in weather suitable for lovers but, instead, three years of hunger, dearth and famine. On 20 February (*2 March*), heavy winds overnight caused extensive property damage, including pushing down trees and throwing a bell from the *Liebfrauenmünster*'s tower into one of the building's doors. Weather during the growing season was too hot and the harvest merely adequate that year. Winter 1534 was unusually cold and an unseasonable frost on 23 April (*3 May*) froze the flowering grapevines, while extreme summer heat created conditions that led to several wooden houses catching fire near Thann and, in Haguenau, dried the wells and fountains. The harvest was poor, leading to dearth in Alsace, while excessive insect life ate the leaves of many deciduous trees around Guebwiller to skeletons. The year finished with more extremely cold temperatures which continued into 1535. That year again, an unseasonable spring frost on 20 March (*30 March*) followed by two weeks of low temperatures damaged the flowering trees, including those carrying apples, almonds, and other nuts. A cold snap in mid-June inhibited the grapes' growth on the Vosges hillsides, while excessive rain after Saint Lawrence's Day (*10/20 August*) rotted the crops, with only ten days of clear skies recorded from that date to Saint Martin's Day (*11/21 November*). Famine followed.[20]

This general pattern of cold winters and hot summers continued; deep cold and a blanket of snow covered the region from Saint Andrew's Day 1535 (*30 Nov/10 Dec*) to Saint Matthias' Day 1536 (*24 Feb/5 March*), leading to a shortage of firewood and a spring of high waters in the region. After a violent rainfall on 15 June (*25 June*), though, which forced a postponement of the procession planned from the Guebwiller church that day, the summer turned hot and dry. Water levels became so low in the rivers that water mills were unable to operate, beasts suffered, and 'men wailed.' Despite the drought, harvests of grain and grape in 1536 were so good that prices were low from Basel to Haguenau. A cold winter in 1537 led into a wet spring and on 23 April (*3 May*) a heavy thunderstorm hit Strasbourg. Lightning caused a fire which destroyed the *Grüne Warth*, a watchtower on the city's southern walls. Rain continued to fall heavily during May and June; with water-soaked soils, a hot and dry summer after the summer solstice led to abundant harvests for 1537.[21]

The string of six very cold winters came to an end in 1538, when temperatures were generally average. An eight-day cold spell in early March froze the soils around Guebwiller; this, along with a six-week drought, retarded the growth of grapevines. Deep frosts hit much of the Upper Rhine Valley in mid-April, freezing young crops and already-delayed vines. The price of wheat soared and, by bringing their stored grain to sell in Guebwiller and Soultz, farmers from the Sundgau, a small southeastern region of Alsace near the Belfort Gap, saved many from starvation. The summer turned hot, with many thunderstorms; destruction from repeated instances of hail and flooding inspired the Dominican chronicler at Guebwiller to write that it was 'a frightening and dangerous year'. In the end, the harvests were good enough, although prices remained high.[22]

The winter of 1539, the first of two extraordinarily productive years for the grape harvests in Alsace, began with colder temperatures. After a surprise frost at the start of March 1539, the weather of April, May, June, July, and August was perfect for farming. The hay was brought in easily and reports from the length of the Upper Rhine Valley – Haguenau, Strasbourg, Guebwiller, Thann, Mulhouse, and Basel – praised the quality and quantity of wine produced that year. Wine barrels were in short supply and began to cost more than the wine they stored. Prices for all the other agricultural products were low, indicating abundant harvests throughout Alsace.[23]

The wine of 1540 was praised as equally exceptional and delicious, but a year of very different weather produced it. Hotter than usual temperatures began in February and, while rain fell for three days at the beginning of March, there was no precipitation in April or May. Some rain fell in the second half of June, and the first harvest of grain was moderately successful. After the solstice, though, summer and early autumn temperatures rose so high that women were reportedly cooking eggs on the cobblestones of Strasbourg. The water level in the Rhine River diminished to such low levels that it could be forded on horseback. Forest fires were a hazard, and both animals and humans died of thirst. The drought became so severe that water was as expensive as wine in several places, according to the chronicler at Guebwiller. Fruit withered on the branch and grapes desiccated on the vine until two days of solid rain at Michaelmas (*29 September/9 October*). This rainfall allowed for an unexpectedly successful grape harvest, leading to a second year of exceptional wine in terms of quantity and quality. It was too late for other farm products, however, and the costs of fruit, vegetables, and grain remained very high.[24]

Alsatians were not alone in facing the extreme weather of 1540. Research by Oliver Wetter and his colleagues shows that the drought of 1540 affected an area of 2–3 million km^2 – the entire European continent from mid-northern Sweden to southern Spain, from England to Romania. From the meteorological perspective, "(t)he spatial distribution of drought and precipitation records indeed suggest(s) that 1540 was dominated by a quasi-persistent high-pressure situation over Western and Central Europe surrounded by low-pressure systems over the Atlantic and Western Russia." Data from Basel, Zürich, Lucerne, and Schaffhausen indicate that in 1540 the number of precipitation days in Switzerland was 81% lower than the twentieth-century average, which is similar to research results for Prague in Czechia. The lack of precipitation contributed to the extreme heat, as soil desiccation is a well-known driver for heat extremes. Overall, the drought and heat of 1540 is considered more extreme than those of 2003 or 2010.[25]

This section provides the final 16 years of the volume's 68-year weather report, from 1525 to 1540, and some of the social consequences of that weather. To summarize: during those years, 11 harvests were reported to be above average (1526, 1527, 1528, 1532, 1536, 1537, and 1539) and average (1525, 1530, 1533, and 1538). During the same period, three harvests were reported to have been poor (1529, 1534, and 1540) and two harvests failed completely (1531 and 1535).

Impoverished urban and rural people in Alsace endured dearth (scarce food supply) twice during these years: in 1530, and 1534. Higher than usual exports to Switzerland in 1530 and numerous refugees from Lorraine (due to weather-related harvest failures in these areas) created dearth in Alsace in 1530, while in 1534 hostile weather in Alsace caused poor local harvests and dearth. The entire region suffered from famine (drastic food shortage) three times from 1525 to 1540: in 1529, 1531, and 1535, all three caused by weather-related harvest failure.

Notably, the social consequences of these periods of dearth and famine did not include plans for rebellion among Alsatian peasants nor calls for it by Strasbourg's poorer citizens, even after three years of famine and dearth from 1529 to 1531. Instead, the city's Reformers, having gained authority over the social welfare system, provided immediate succour to some refugees arriving at the gates, sent others to work for the city, and refused any long-term responsibility for all of them.

Thus concludes this report on weather events in the Upper Rhine Valley from 1473 to 1540 inclusive: 68 years covered over three chapters and bookended by two years of extreme drought and heat. The weather in Alsace over the following decades continued to be difficult for Alsatians as the Spörer Minimum gave way to the Maunder Minimum of the Little Ice Age. However, the focus of this research is on the diversity of religious views of the natural world and these were largely established by 1540; a longer weather report is outside the remit of this volume.

Case Study 3: 'In Principio:' the natural world in *Hexemeron Dei opus*

The third case study of this volume turns to the early Evangelical community in Strasbourg for another view of the natural world with an exploration of Wolfgang Capito's *Hexemeron Dei opus*. Following an introduction of the work, its publisher, its influence, and its context, the case study will report and discuss Capito's exegesis with a focus on his representation of the natural world and his views of the appropriate Christian relationship to it.

With almost 600 pages of exegesis focused on Genesis books 1 to 11, as well as a dedication, preface, and other features, *Hexemeron Dei opus* was Wolfgang Capito's last written work. It was published in 1539 by a friend and fellow scholar from Hagenau, Wendelin Rihel, and there is evidence of only a single print run. Rihel had received municipal citizenship from Strasbourg in 1525 and by 1531 was known to be operating a bookshop; by 1535, the operation had expanded to include printing. Chrisman's research shows him to have operated one of the larger establishments in the city, with two presses working at capacity for 16 of the next 20 years. A search through the Universal Short Title Catalogue shows several works by Martin Bucer, Martin Luther, Petrus Dasypodius, Hieronymus Bock, Aristotle, and Cicero among his editions, as well as those by Wolfgang Capito and Jean Calvin.[26]

There are several reasons to believe that Capito's *Hexemeron Dei opus* was included among the texts studied by the pupils of Johannes Sturm's *Schola Argentoratensis*, including the date of publication (it was printed in 1539, the year following the school's beginning) and the fact that it was published by Rihel, whose main focus was on printing Protestant and classical texts for the new school, located conveniently nearby. Capito was deeply involved in the work of founding the institution, which united his interests in education as both a reformer and a humanist, and he gave public lectures at the *Schola Argentoratensis* on the Old Testament. The language of the text is Latin, the language used to teach at the school, with Hebrew scattered regularly throughout. When the *Schola*'s statutes were revised in 1545, a greater emphasis was placed on theological education, with lectures to be focused specifically on, among other named books, Genesis. Upon Capito's death, the only biography from a living peer was Sturm's brief description in a volume of student exercises published in 1542. However, although Capito was clearly closely involved with the *Schola Argentoratensis* and friends with its rector, and although the style and subject matter of the *Hexemeron Dei opus* were appropriate for student study, there is no documentary evidence listing the book as part of the curriculum for the school's pupils.[27]

Nevertheless, the weight of circumstantial evidence is highly suggestive that Hexemeron Dei opus was used as a text. If so, Capito's exegesis on the first book of Genesis would have exerted no small degree of influence on the manner in which the sons of the elite came to understand the natural world.

The Strasbourg reformer was not, of course, the first theologian to reflect upon the creation of the world as represented in Scripture. He published amidst a flurry of commentaries by early reformers in southwestern Germany and northern Switzerland, including Zwingli (Zurich, 1527), Konrad Pellikan (Zurich, 1532), Johannes Oecolampadius (Basel, 1536), Peter Martyr Vermigli (lectures delivered in the early 1540s, published posthumously in 1569), Luther (Wittenberg, 1545), Calvin (Geneva, 1554), and Wolfgang Musculus (Bern, 1554).[28]

Hexemeron Dei opus is a large volume, being some 650 pages published as a quarto. It begins with a 14-page preface dedicating the work to Duke William of Cleves (1516–1592), followed by a substantial index of 16 pages. The pages are double-sided, with the text displaced towards the binding, which allows ample room for the many marginalia with which the book is liberally sprinkled. After the index are ten pages presenting the 31 verses of the first book of Genesis, with snippets of Torah commentary from the Targum Onkelos and the Jerusalem Targum; both of these are Hebrew commentaries on the first book of the Torah, translated by Capito into Latin. The first is considered today to be an eastern or Babylonian edition, while the second is understood to be a western or Israeli text. Scripture is offered in Latin with large font, widely spaced, while the comments are smaller and cramped, with Hebrew and Greek terms included. The large volume mentions the natural environment frequently; the following passages were selected to facilitate a comparison with Geiler and Clemens Zyegler, authors featured in the other case studies in this volume.

Following Capito's final words of justification for the book, he articulates the conceptual framework for his work about Creation: any reader with confidence in Scripture who should contemplate the ordering of the world will quickly recognize that God's strength and power are infinite, and that the universe and all its parts depend upon God's will and word. That is, once persuaded that everything was created *ex nihilo* by the Father's co-eternal Word, one concludes that this one and the same Father keeps all things in existence as long as he wishes. Whether pious or impious, Capito continues, the reader will realize at once that power does not give rise to itself, nor does the hand of change move itself. The perception simultaneously arises that we are part of the universe, which, in its entirety, arises and abides through the power of the one Word of God, and until it should pass away, He acts and rests by turns. The reader will recognize that all things which have been made are the prerogative of him who created them. In this way, Capito writes, God governs everything by a nod or shake of His head, just as He created all things by His word, and there is no one who would hinder Him.[29]

Capito makes no attempt here to explain the manner in which God created the material world or the manner in which He sustains it, nor did he later offer explanations for these. Capito's God is simply simultaneously and intimately present to all of Creation and unknowably mysterious; God's direct attention is responsible for everything that exists and everything that happens. Presumably, wilderness and the activities of wild animals are included. There is no distance or distinction between Creation and God's will; according to Capito, to learn about Creation, therefore, is implicitly to learn about God's will for it.

Having asserted his view of Creation and God's will as inextricably inter-woven, Capito goes on to establish Scripture as the only appropriate manner of understanding both. His conflation of God's will with the natural environment and his insistence that Scripture is the sole interpretative filter means that other approaches to understanding natural events or processes should be disregarded. This includes, for Capito, the methods bequeathed from Classical philosophers and written material from other authors who contemplate the natural environment without the illumination of the Holy Spirit, both of which are derisively dismissed as 'watchtower knowledge.' For Capito, this sneer means knowledge which is derived from watching Creation (God's Will) when God is known, but not for the purpose of glorifying Him; such knowledge is inadequate because it does not, in Capito's opinion, incline a person towards doing good.[30]

Capito's position was not unusual for an Evangelical reformer. For example, Luther's objections to Aristotelian philosophy were central to his critique of the Roman Catholic Church and were expressed forcefully and repeatedly from February 1517 onward – even prior to his attack on indulgences. The core of the objection to Aristotle was that the Greek's philosophy was the central pillar upon which hung the conviction of medieval scholastic theologians that theological truths were attainable through human reason. This was, in Luther's view, a profoundly mistaken approach, and this led him to conclude that natural philosophy, as found in the early sixteenth-century university, was actively endangering hopes of eternal salvation. As Sachiko Kusukawa points out in her 1995 monograph *The Transformation of Natural Philosophy: The Case of Philip Melanchthon*, "Luther's attack on Aristotle (…) was thus an attack on the whole basis on which scholastic theology rested."[31] In his 1520 tract *Address to the Christian Nobility of the German Nation*, Luther introduced his programme for university reform: the faculty of arts – through which every undergraduate student continued to matriculate – should retain Aristotle's *Logica*, *Rhetorica*, and *Poetica* for their uses in preaching and speaking, but his philosophical books (*Physica*, *Metaphysica*, *De anima*, and *Ethica*) should be abandoned. This disagreement with natural philosophy, however, was based on the use made by medieval theologians of the method of reasoning proposed by Classical thinkers; both Luther and Philip Melanchthon came to accept the need for reliable knowledge of the natural world, particularly for medical purposes. This topic will be explored in greater depth in the next section.[32]

Denying the validity of Aristotelian-influenced natural philosophy as a useful method for developing an understanding of the natural world had intellectual consequences from which Capito does not shy away. He advises a studied indifference to the reasons for essential differences or unique phenomena, such as those between regional rainfalls or life spans; knowledge about such things, he claims, was purposely withheld by God and not for humanity to share. Therefore, Capito asserts, it was not for humans to know why one person has greater gifts than another, nor even why, of two people in one bed, one is taken and the other left. Uncertainty, he claims, is fitting to our finite human nature, and revelation only sets forth that which builds people up in God through Christ. Humans do not know why God would have provided such an abundance of different things, nor the reasons why some were made available for contemplation at certain times, places, or circumstances. Capito tells his readers that even if their admiration convinced them of the surety of God's will, labouring to know more about what lay before their eyes or anything the creator was not pleased to disclose through Scripture is not to be undertaken.[33]

Along with the prohibition on Scripturally-uninformed investigations of the natural environment, Capito regularly asserts that the very same natural world was created solely for the sake of human beings. The opinion that humanity was given full authority over the natural environment by God frequently accompanies this assertion, such as during his exposition of Genesis 1:28 when he articulates his view of the position of humanity in the hierarchy of Creation. Unsurprisingly, that position is one of dominion. Capito expands upon the simple clause, however, to extend this God-given authority into the fruits of human industry, which thereby renders the whole world as subservient to man through energy and labour. To clarify, Capito claims that when humans plow, log, fish, or mine, they are fulfilling God's command to exert dominion over the earth. He goes on to write of the earth's subjugation as marvellous due to the crops produced by plowing, harvesting, and harrowing, and the minerals produced by mining. These allow men to live comfortably and pleasantly, proving, according to him, that on earth man alone rules. He interprets Psalm 8 as deriving from this point, pointing out that God set man over His works and subjected all things under his feet.[34]

How does Capito explain humanity's privileged position in Creation? Founded in his exegesis of Genesis 1:26, Capito's explanation for humanity's powerful position is that it is due to having been created in the image of God. An effect of this is humanity's capacity to rule over animals, because external activity and interacting with other beings follow from one's substance and implanted character. Adam, he writes, was made in God's image, and that image included God's authority over the fish, birds, and animals. Just as man resembles God in goodness, wisdom, and righteousness, it is also fitting for him to obtain the faculty of divine authority in governing the world, in order that he might develop a similar consciousness.[35]

Capito tidily slots Adam at the pinnacle of Creation, obedient to God and authoritative over the animals, birds, and fish. Creation's hierarchy is emphasized by Capito's ascription of 'royal authority' to God, extending the conceptual metaphor which organized his view of the world. A well-developed and intricate metaphor by the early sixteenth century, terms for God such as

'Lord' or 'King,' which were also used for the highest civil authorities, were found throughout Christian faith and practice. Capito's (and the reformers') innovation was in the immediacy with which they presented that authority to human beings, without mediation from saints, angels, or other spiritual entities. The importance of this conceptual metaphor for understanding Capito's view of Creation is emphasized by the punishment earned by Adam for disobeying God and eating the fruit of the tree of knowledge of good and evil. He writes that filthy ignorance was present in place of knowledge, hatred in the place of affection, rebellion instead of obedience to God and instead of orderliness, there was chaos, irreconcilable strife, and outrage.[36]

In a social hierarchy whose stability required acquiescence to the wishes of superiors, disobedience (such as that displayed by the peasants more than a decade earlier) was a primary offence, and Capito speaks fulsomely and repeatedly about the severe consequences of disobedience to God. It is not too large a stretch to speculate that his emphasis on obedience as a pre-eminent Christian virtue has as much to do with his desire for civil order as it did an individual's eternal salvation and shows his personal deference to the desires of Strasbourg's civil authorities. For the adolescents who were likely reading *Hexemeron Dei opus*, and who, if all went well, could occupy positions at the upper end of the social order, each reminder of the value of obedience may have served to reinforce what was due to them from social inferiors, as well as what was due from them to God.

Alongside his central desire to represent Creation as ordered and ranked, Capito does provide a few explanations for natural processes which credit both natural causes and divine fiat. In explaining, for example, the germination and growth of plants, the reformer's exegesis on the clause *Germinet terra germen* (Genesis 1:11) allots generative power to the earth itself, mediated by seasons and the weather. He also mentions that some think plants sprout 300 times more abundantly from ground beneath which is a buried human cadaver, because when the human body returns to earth, it fully represents man's native condition.[37]

The reformer takes advantage of God's creation of animals on the sixth day to elaborate on the differences between them and human beings. He does this by refining upon distinctions between two types of soul – the living soul, found in animals, and the immortal soul, bestowed upon humans. A living soul, such as that found in reptiles or mice, is a body endowed with life (vitality and the principle of movement). But the immortal soul of a human being requires a body and members to do the work of breath and respiration; he positions the human soul between the boundary of the body and the breath of life. The soul and spirit of humans, which, he writes, are called the same thing, differ from the souls of animals because the souls of animals are made from matter and are corruptible, while the soul of humans is beyond matter and was gifted to men by God. Another passage on animals focuses on the fear of mankind shown by wild animals, which is interpreted as proof of the authority given by God to man over them. The superiority of man is ascribed not to his natural qualities, strength, or mind, but to the power of the divine blessing which was bestowed upon Noah's descendants at the Flood and which endures even in corrupt sinners. This passage, like many others, illustrates the *a posteriori* nature of Capito's exegesis, in that he provides knowledge which proceeds from observations or experiences to the deduction of probable causes. That is, Capito provides empirical observations of the natural environment which are taken as proof that God's will is the primary cause for the nature of those observations.[38]

Capito's exegesis of Genesis Book One scrupulously avoids the Aristotelian natural philosophy with which, as a university-trained theologian, he would have been intimately familiar. Nonetheless, in keeping with the demand from Strasbourg's *Rat* to foster obedience and order in the Reform community, he draws upon familiar resources to frame and articulate his view of God's work in making Creation. These resources were primarily his skills with ancient languages, allowing comments from the Targum Onkelos and the Jerusalem Targum to illuminate the *Hexemeron*, but also the concept of the Great Chain of Being. The idea is obviously operative in Capito's exegesis, but in a sharply abbreviated format: humanity is no longer positioned somewhere in the middle, below the mediating angels, saints, devils, or demons, but immediately beneath God. The animals, as well as plants and minerals, continue to occupy subordinate positions to men and women.[39]

To summarize this case study, there are two primary observations which may be offered about Capito's representation of the natural environment in *Hexemeron Dei opus*, leading to two conclusions and several more questions. The first observation of the work is that Wolfgang Capito explicitly and repeatedly articulates the centrality of humanity to Creation's purpose, allocating a privileged place to human beings in relation to all of the created sphere, including the natural world. It is true that Capito's subject is Book One of Genesis, during which God creates the earth and all the creatures on it, so it may be suggested that a central role for humanity is a reasonable expectation for an exegesis of that Book. However, Capito's understanding of the full union of God's will with the natural world builds on his refusal to integrate natural philosophy into his exegesis, leaving only one way to formulate an appropriately Christian understanding of events, forces, and processes in the environment: as direct messages from God to human beings.

Moreover, given that one of the roles of Genesis in the Bible is to articulate cosmogony from a Judeo-Christian perspective, Capito's comments upon Book One re-interpret an indispensable and central text from the Reformed perspective. Culturally, his work contributed towards solidifying the intellectual and religious foundations of the new society the Evangelicals were building in Strasbourg and elsewhere. Future research into the many theological differences between Catholic, radical, and Protestant exegetes may benefit from an in-depth comparison of pre- and post-reform commentaries on Genesis, but that is not the purpose of this work.

In the context of religious perspectives on the natural world available to Strasbourgeois from 1509 to 1540, *Hexemeron Opus dei* stands out for the manner in which Capito starkly asserted that Creation was made for human beings to exploit. It is difficult to avoid the second observation: that, aside from God's intentions for humanity, Capito's work allows no purpose for the existence of Creation or any creature other than humans. Although his perspective did not emerge *ex nihilo*, nearly 500 years of familiarity with it may impede our appreciation of the unwonted degree of exploitability ascribed to the natural environment in Capito's exegesis, where strident emphasis was placed on its primary purpose as a platform for human destiny.

Capito's views of nature in the context of the wider Evangelical movement

The period from 1530 onwards was a time when unity among the Evangelicals was consolidating around the views of Luther and Melanchthon, making it useful to situate *Hexemeron Dei opus* among the views of other reformers. As the reformers gained influence, universities which were sympathetic to the reform changed their curriculum in order to structure it similarly to that of the institution at Wittenberg where, as early as 1526, the undergraduate Arts degree was re-oriented towards Latin grammar, dialectic, and the elements of rhetoric and of mathematics. The languages, rhetoric, and dialectic were all seen as essential for handling theological questions confidently; rhetoric, in particular, was positioned as crucial to the development of the eloquence needed by preachers. Achieving a Master of Arts required knowledge of the Greek language, through which knowledge of the natural world was gained from Classical authors; knowledge of plants, gems, and living beings was particularly advised, due to their importance for medical issues. Advanced study also included more rhetoric, the works of ancient orators, and advanced mathematics.[40]

Notwithstanding their criticisms of scholastic natural philosophy, some reformers turned anew to Aristotle, initially in order to find an intellectual foundation for reforms in the legal arena. By 1532, Philip Melanchthon found himself using the *Ethica* once more, this time as the basis for teaching the moral philosophy of civil obedience. Carefully distinguishing his use of the *Ethica* in demonstrating precepts from the 'law of nature' (necessary for orderly civil life) from the role it played for the Catholic Church in the theology of justification or of good works, Melanchthon presented moral philosophy as part of Divine Law, although not included in the Gospel. The law of nature was often defined according to Romans 2:13–15, interpreted as natural inclination towards goodness, and had developed through the Catholic jurist tradition to mean the instincts divinely bestowed upon all humans, along with others such as that of self-preservation. Natural law frequently served as justification for civil law and furnished the basis for the political philosophy of both Melanchthon and Luther.[41]

Despite returning to Aristotle for philosophical support in achieving the civil aspects of reform, Melanchthon was meticulous in insisting that natural philosophy, as part of Divine Law, must be known in the context of the Gospel. For example, in a poem from 1532, he praised God the Creator who placed in man the law of nature for civil virtue, which the *Ethica* usefully teaches; Christ, however, was necessary for the Word of God to live in man. *Physicae seu naturalis philosophiae compendium* (1543) and *Initia doctrinae physicae dictata in Academia Vitebergensi* (1549), two later works of natural philosophy authored by Melanchthon, did not waver from this perspective.

In Sachiko Kusukawa's 1995 exploration of Melancthon's thought on natural philosophy, *The Transformation of Natural Philosophy: The Case of Philip Melancthon*, the three books of the *Initia doctrinae physicae* show Melanchthon's understanding of natural philosophy as sharing many features with Capito's *Hexemeron*, including a view of the world as having been created for the sake of man. However, Melanchthon is primarily interested in the "consideration of matter, of the qualities in matter and their effects, which are the causes of changes in the bodies, such as generation, nutrition, alternation and corruption."[42] In the first book, he justifies the validity of a natural knowledge of the existence of God, defines Providence as "knowledge by which God foresees everything and a government by which He protects His whole creation,"[43] and deals with the motions and effects of the sun, the moon, and the planets. The second book features an exploration of the kinds and classification of causes; for example, an Aristotelian argument eliminating accidental causes as due to fortune or chance is illustrated with Scriptural examples. The argument concludes that "because it is impossible for this most beautiful order of things (...) to have come into being by chance or to subsist merely by chance," the material existence of the world is a "pointer to God."[44] The third book examines the four elements and the causes of various changes in matter. As Kusukawa notes, the message of the work is "to demonstrate God's Providence throughout the physical world in an Aristotelian way."[45]

In the last book, as Kusukawa comments, Melanchthon has come to consider natural philosophy worthy in its own right. The whole nature of things, he writes, is like a theatre for human minds which God wished us to view and for that reason he placed in our minds the desire for considering things and the pleasures which accompany gaining knowledge about them. These reasons, according to Melanchthon, invite healthy minds to the consideration of nature, even if not for utilitarian purposes; the mind is led to beholding things by its own nature. To consider nature, then, writes Melanchthon, is to follow one's own nature and it leads to the most pleasant joy.[46]

Capito was acquainted with Melanchthon and Luther, as mentioned earlier, but he was not a partner in Melanchthon's renovations to natural philosophy. Although the *Hexemeron Dei opus* was an extensive exploration of God's creation of the natural environment, Capito was unequivocally working with the Gospel and not exploring the natural world for medical purposes or for its own sake. Capito's constitution of the natural world as having been created for humanity is not a prohibition against observing and learning about it, as demonstrated through his exuberant appreciation of improvements in agriculture and mining as fulfilling God's order to subdue the earth. However, Capito is primarily interested in inspiring his readers to have greater faith in God with the *Hexemeron*. Having dismissed natural philosophy as 'watchtower knowledge,' Capito asserts that revealed knowledge (that is, Holy Scripture) is eminently practical. This knowledge, he writes, comes into play not so much in one's thoughts as in one's behaviour, where it is tireless, and unceasingly concerned with entrusting ourselves to God, calling upon him, accepting all things from him, and giving thanks to him alone. As a highly regarded theologian exercising his craft, Capito focused his final publication on the meaning of Creation for Christians, on the relevance of Genesis' Book One for human salvation, and on God's purpose for making the world the way it was – not 'the most pleasant joy.'[47]

Capito can be understood as subsuming an intellectual knowledge of the natural world into his theological vision of Creation, while Melanchthon accepts such knowledge as valuable for its own sake and for human purposes. Could the popular desire for practical means of understanding the environment (as exemplified by the best-selling volumes *Wetter Büchlin* and *Bauernpraktik*) have contributed to the latter's motivation for developing and maintaining his perspective? Melanchthon, and Luther with him, were careful to emphasize the medical applications of Aristotelian-based knowledge, and there is no evidence suggesting that Reynmann's *Wetter Büchlin* or the *Bauernpraktik* were considered by them as relevant to salvation or perdition. Nevertheless, their understanding of their culture was sophisticated and nuanced, and such an accommodation to the practical needs of their new and potential converts is not beyond the realm of possibility.

Did these two theologians, Capito and Melanchthon, straddle the historical divide between the unity of science and religion and their development into distinct fields of knowledge? Neither author would accept the pure materialism of modern natural sciences, but its roots are visible in *Initia doctrinae physicae* whereas they are not in *Hexemeron Dei opus*.

Jean Calvin and nature in *Institutio Christianae Religionis* (1539 edition)

Born in Picardy on 10 July 1509 – a few months after the death of Johannes Geiler – Jean Calvin showed early dedication to the Roman Catholic Church: by the age of 12 he had received a tonsure and was employed by Charles de Hengest, Bishop and Count of Noyon, as a clerk. His father chose that he should pursue education to become a lawyer and by 1532, the young man had earned his licentiate in law from the Université d'Orléans. Calvin's conversion from Catholic to Reformist thinking is understood to have taken place sometime during the second half of 1533. Suspected of involvement in preparing the text of Nicolas Cop's infamous inaugural address as the new rector of the Université de Paris, which, according to Gordon, featured 'an Erasmian account of scripture with unmistakably Lutheran overtones,' Calvin fled Paris in November 1533, After the Placards Affair brought swift retribution to French Reformers from Francis I, the young lawyer left France for Basel, Switzerland in October 1534.[48]

By the autumn of 1535, Calvin had completed his first edition of *Christianae Religionis Institutio* (Of the Christian Religion, an Institution), although it was not printed until the spring of 1536. After a visit with Renée of France at her court in Ferrara, Italy, and a quick trip to Paris on family business, open conflict between the King of France, Francis I, and the Holy Roman Emperor, Charles V, blocked the usual route back to Basel along the Rhine River and inspired Calvin to travel through Geneva, newly assumed by the city of Bern into its territories. There, the inflammatory preacher Guillaume Farel, a fellow Frenchman, confronted Calvin's choice to devote himself to private studies and, with threats of God's wrath, seems to have bullied him into serving the Evangelical church (although Calvin claims to have felt the hand of Providence in the decision). Calvin's growing prominence in Geneva – he was delivering daily lectures on the New Testament, had published new material since the Institutes, and had established relationships with major reformers of the region, including Bucer, Capito, Megander, Myconius, and Bullinger – secured his position as a leader of the Reform in that city. Explicitly intended to act as the foundation of the municipal church, Calvin's *The Confession of Faith* (1536), the *Articles Concerning the Organization of the Church* (1537), and the *Catechism or Instruction of the Christian Religion of the Church of Geneva* (1537) articulated the standards of orthodoxy in newly reformed Geneva; refusal to obey them would mean excommunication from the church and exile from the city. Calvin was unwilling to accept any political supervision of the Genevan church and when the Bernese insisted that their liturgical rites be practised in Geneva at Easter 1538, Calvin and Farel defied the municipal government (who accepted the dictum) and departed the city.

At the urging of Bucer and Capito, Calvin settled in Strasbourg and received citizenship in July 1539. Their aim was to teach him to be a pastor for the 400 or so French refugees in the *Freie Reichsstadt* and to organize them into a parish. In September 1538, he began to preach in the church today known as Saint-Nicolas-des-Ondes and, soon, in the chapel of the penitents at Sainte-Magdelène (formerly a convent for reformed prostitutes). Integrating Bucer's Strasbourg liturgy with its emphasis on congregational singing, Calvin eventually preached or lectured every day (twice on Sundays) while he was in Strasbourg. Taking on a function denied to him in Geneva, Calvin also began to offer pastoral ministry to individuals, with some success not only in providing consolation but also in inspiring Anabaptists such as Hermann of Liège to join the Reformed church.

Bucer and the Strasbourg Reformers' important role in the maturation of Calvin's theological views, pastoral abilities, and political acumen is unquestioned in historical research today. On the purely practical level, Bucer initially hosted him in Strasbourg, found employment for him, and, eventually, found a house for the Frenchman that shared a back garden with his own so the two could meet frequently. He brought Calvin into his circle of intimate friends and through Latin, their shared language (Calvin never learned German or Alsatian), the two gave many evenings to deep conversation. Along with Hedio, Calvin provided commentary on the New Testament at Johannes Sturm's *Schola Argentoratensis*, while Capito and Bucer lectured on the Old Testament. In the summer of 1940, after at least four matchmaking attempts by his new friends, Calvin married Idelette de Bure, the widow of an Anabaptist and the mother of two children; Farel came to Strasbourg to bless their union.

During his stay in Strasbourg, Bucer, Capito, and Sturm mentored the French Reformer in the skills needed to lead a nascent church, although Calvin reportedly had no intention of returning to Geneva at the time. They brought Calvin into the wider world of imperial religious politics, including him, for example, among Strasbourg's representatives to the Leipzig Disputation in January 1539, the imperial diet in Frankfurt six weeks later, and a meeting of the Schmalkaldic League two months after that, where Calvin met Philip Melanchthon. Gordon considers that under the influence of these three prominent men, along with his growing personal acquaintance with Melanchthon, Calvin began to turn away from Zwinglian theological views and towards Lutheranism. This culminated with Calvin adding his signature to Melanchthon's 1540 *Confessio Augustana Variata*, a revision to the original Wittenberg Concord of 1530. The Frenchman went on to represent Strasbourg's interests in the religious colloquies of 1540 and 1541 where Catholics and Reformers attempted to find unity; during the latter, Calvin joined the party of the Duke of Lüneberg, a Lutheran Prince, with Sturm.

While honing his skills, Calvin also witnessed Strasbourg's Reformers struggle with the issue of discipline of the faithful: essentially, whose right was it to excommunicate someone from the church? At issue for the Reformers was addressing disaffection from worship in the face of the new doctrines, as well as improving the moral life of the city and protecting the Eucharist from unworthies. However, having accepted the responsibility for appointing and paying ministers, maintaining church buildings, and other leadership roles formerly exercised by the Roman Catholic Church, Strasbourg's *Rat* insisted that only they had the authority to excommunicate members of their city's churches. In the end, Bucer, Capito, Hedio, and Zell, having enlisted the *Rat*'s assistance in dislodging the Roman Catholic Church from the city in the 1520s, failed to regain authority over a disciplinary system independent of the magistrates in the 1530s; instead, the *Rat* decided to appoint three disciplining *Kirchenpfleger* (elders) to each parish and to retain the power to excommunicate. Cheek by jowl with these efforts, meanwhile, Calvin controlled admission to communion in his French church of refugees by only accepting those who had presented themselves to ministers for examination.

In addition to learning leadership skills and exercising pastoral responsibilities, Calvin continued to write. Published in the same year and by the same Strasbourg publisher as Capito's *Hexemeron Dei opus* (Wendelin Rihel), Calvin's 1539 edition of *Institutio Christianae Religionis* – the second of five Latin editions and six very similar French translations from his pen – saw the introduction of several new topics. The volume doubled in size from its first edition, from 6 to 17 chapters, with pages now numbering over 400. The title of the book was modified and subjects now included the universal fact of religion, free choice, church offices, human traditions, excommunication and the civil magistry, as well as the introduction of Calvin's contentious ideas about election and predestination. According to these, as Gordon summarizes, God elects the faithful before they are born; without election there is no salvation; human merit has nothing to do with election; and God never abandons the elect.[49] The new edition was no longer a catechism but had become a book of instruction aimed mostly at ministers of the Reformed church, who were to read the Institutio and his commentaries as an introduction to Scripture. Calvin's comments in the Dedication reveal his view of his own role in the Reform: where Bucer saw negotiation and compromise as the avenue to unity among Reformers, Calvin felt that as long as a primary commitment to Scripture was present, doctrinal differences could be accepted and he, himself, was uniquely skilled and called to comment on Scripture. Based on his capacity to interpret and comment upon Scripture, then, the Frenchman saw his role as that of mediator between various Reformist churches. From 1539 forward, Calvin intended the *Institutio* to serve as a guide to the entire Christian religion, with Scriptural references and commentary organized thematically.[50]

Calvin's focus in the 1539 edition of *Institutio Christianae Religionis* was so exclusively on the salvation of the Christian soul that there is very little explicit mention of the natural environment outside of its function as an inspiration for human contemplation of God and as a vehicle of God's providence. For example, early in the volume Calvin describes extreme weather events such as sky-shaking thunder, lightning which sets the air afire, and the turbulence of the sea followed by calm, and after ascribing them clearly to the direct agency of God, writes that they are to serve as reminders of God for those who witness them. A few pages later, the author asserts that unusual natural events are driven by the providence of God, not by blind fortune. Partway through the second chapter, Calvin makes a distinction between earthly things and heavenly things, but the group of earthly things includes only politics, economics, all the mechanical arts and the liberal disciplines. These are all human activities; the natural environment of earth itself is not mentioned. In a very few comments, then, Calvin begins to introduce his view that the earth was created to inspire humanity's contemplation of and devotion to God, whether in awe of the creator or by accepting God's will through natural events.[51]

In a section in the fourth chapter focused on the creation of heaven and earth, Calvin speaks more overtly to the active role of God in natural occurrences. He asserts that the world is clothed with the word of God, its creator and constant sustainer, and illustrates the intimacy of this relationship when writing that it is by the hand of God that dew and rain fall to irrigate the ground; that God supplies food to the chicks of crows; and that a sparrow does not fall to the ground without God knowing and so disposing of it. While a pious soul may take comfort from the unceasing support of a powerful divinity, Calvin writes, the indignation of God about the sins of human beings may be seen in barrenness, famine, and pestilence. Despite the title, the matter of the section once again serves for Calvin to call Christian souls more powerfully to worship God in a suitable manner (that is, according to Calvin's own doctrines) by reminding them of God's power to give and withhold material well-being through natural events. That is, God works through nature to reward piety or punish sin. There are few further mentions of the natural world in the book and those that are present express these two ideas.[52]

As noted by Susan Schreiner in *The Theatre of His Glory*, Calvin did not recant his views on creation in further editions of the *Institutio*, but rather developed them in greater depth. While there are only a few small mentions of the topic in the 1539 edition, by the final 1555 edition Calvin's views that the cosmos is a reflection of God's glory, and that God's active providence works through nature, were clearly discernable, along with a deepening discussion of the relationship between natural law and society. Schreiner's discussion concerning the final evolution of Calvin's thought about nature is instructive; in brief, according to the Reformer, creation fell with Adam and Eve. This leaves the natural world hanging between chaos and harmony and requiring that God hold creation constantly in existence; this makes any independence from God impossible without immediate collapse. God's active and immediate will serves not only to explain the existence of the world but also to describe the restraint of disorder. In adopting this perspective, Schreiner explains, Calvin sought to affirm God's active presence in the world and to counter the notion of a distant, clock-maker God whose involvement in the world decreased after the act of creation.[53]

Calvin only lived in Strasbourg for three years. While these were three pivotal years for him personally, pastorally, and theologically, the brevity of his stay has led to only a few comments here about his theological representation of the natural world. Calvin cannot be considered a member of Alsatian culture and, as a short-term resident, he was unlikely to have been deeply affected by the impact of extreme weather events of the Spörer Minimum upon Strasbourg and Alsace, particularly since he remained linguistically and professionally a participant in French culture. Geneva's magistrates began to press for Calvin's return and, although the Strasbourgeois were reluctant to release him, in September 1541 a mounted escort accompanied him from the Rhine Valley to Geneva. (Idelette, the children, and their household goods travelled later by carriage.) While his career in Geneva is considered the most important legacy of the French Reformer, details of that period will not be included here; this volume is centred on Strasbourg and Alsace, so as Calvin departed the Upper Rhine Valley in 1541, he departs also from this study.[54]

Johannes Sturm's *Schola Argentoratensis* and the influence of Reformed views of nature

Strasbourg's Protestant leaders were keenly aware that external religious conformity under duress was an inadequate foundation for the Godly city they intended to establish (although they eventually imposed conformity anyway); a vital component of their ideal reformation was an educated populace who would be able to read the Gospel for themselves and thereby find conviction. Along with the welfare reform described above, the Evangelical clergy urged civil authorities in 1524 and 1525 to assume responsibility for education, and with notable success: on 9 February 1526 (*19 February*), a School Board was established as a permanent commission of the *Rat*. With three members (one noble, two common, all serving for life), the Board assumed control over existing schools. The members went on to establish a library; found lectureships in Hebrew, Greek, and rhetoric; create a school treasury from the properties and incomes of former Dominican, Franciscan, and Augustinian convents; and redirect benefices from St. Thomas to the support of teachers. A college for Protestant preachers was established in 1534. By 1535 there were nine elementary schools for boys and six for girls, and 1538 saw the establishment of the *Schola Argentoratensis*, an advanced Latin school for the sons of the elite which began to accept students at Easter 1539.[55]

Under founding rector Johannes Sturm, the *Schola Argentoratensis* acquired "an international reputation for general, not professional, education of a superior kind, and its aims and clientele expanded far beyond Strasbourg's province, region, or political orbit to make it a model of humanist pedagogy for much of the German-speaking world."[56] A faculty of ten teachers and eleven or twelve professors was assembled to instruct the youth, among whom mingled the sons of wealthy burghers and young noblemen from West Central Germany, Hungary, Bohemia, Poland, and Prussia. Boys entered the Schola at the age of five and remained for nine years. A report from Sturm's chief of staff, Petrus Dasypodius, in 1556 articulates the general curriculum as it had developed in the sixteenth century: the first three years were devoted to learning the Latin accidence, supported with Cicero's *Letters*; the fourth year saw the introduction of Greek, with Latin syntax building on previous instruction. Years Five and Six made a study of prosody, illustrated by Virgil's *Eclogues* and Cicero's *De Senectute* and *De Amicitia*, as well as two of Aesop's Fables in Greek. Three of Cicero's orations and the *Aeniad I* and *II* were the focus of Year Seven reading, while the more able pupils wrote verses. The final years featured the study of dialectic, with readings from Virgil, Homer, and Plato. Religious instruction, given on Sunday, consisted of catechism in the younger years and the Gospels and Epistles for the older boys. Early lecturers at Sturm's school in Strasbourg included Martin Bucer, Caspar Hedio, Paul Fagius, Peter Martyr Vermigli, and, notably for the purposes of this chapter, Wolfgang Capito, who lectured on the Old Testament, and Jean Calvin, who taught New Testament interpretation.[57]

In the context of the *Schola Argentoratensis*, Capito and Calvin's views of nature as a metaphorical glove for God's open hands would have significantly greater influence on European culture than, for comparison's sake, Zyegler's view of the natural environment. Sturm's school fostered the spread of Evangelical ideas among the elite, as it was intended to do, and that the commonly held Reformed view of the natural world as existing only for human purposes became a dominant European perspective is obvious.

Conclusions

The two volumes explored in this chapter – Wolfgang Capito's *Hexemeron Dei opus* and, more briefly, Jean Calvin's 1539 edition of *Institutio Christianae Religionis* – represent the natural environment with two new perspectives in common: first, that God is the exclusive and immanent actor in a natural world which was created for humanity's use and as a platform for our salvation, and, second, that due to God's immanence, natural events exclusively reflect God's will, such that beneficial events are rewards for human piety and destructive events are punishments for human sin. Along with removing spiritual agents (i.e. demons, saints, or the Blessed Virgin Mary) as mediators between humanity and God, Evangelical theology discredited ritual practices such as rogations, cloud-blessings, or ringing the bells as viable means of protection from catastrophe of any sort. As mentioned in the conclusions to Chapter 3, the disappearance of these practices from the public domain could be attributed as easily to radical religious thinkers as to Evangelicals; it is difficult to discern which might hold greater responsibility for the end of rogations, cloud-blessings, or protective bell-ringing in Strasbourg. However, it is certain that the *Rat*'s ban on rogations and other ceremonies in 1529 was simply a formal gesture ratifying a theological and cultural *fait accompli*.

There is no question that a variety of Christian sects struggled for survival in Strasbourg after the 1525 Peasants' War. That the Evangelical view became the dominant and, then, official position of the city's *Rat* may be explained by several factors, one of which was the Evangelical vision for new methods of addressing the situation of indigent citizens and refugees to the city. From Capito's perspective, as expressed in *Hexemeron Dei opus*, good Christians should no longer accept the idea of demonic forces, witches, or most of the other actors identified by Catholic preacher Johannes Geiler as causing stress from the natural sphere; instead, God's will was presented as directly responsible for all such events. By removing other supernatural actors from agency over nature, Capito, Calvin, and fellow Evangelicals left few options for succour except prayer directly to God and human compassion. As a result of this religious change, certain social initiatives were justified and even mandated by the newly dominant perspective, and in several southwestern German cities these initiatives included a complete restructuring of the cities' welfare systems. Over a relatively few years, Evangelicals in Strasbourg succeeded in removing control of the city's welfare system ('poor relief') from Roman Catholic institutions and inspiring the *Rat* to establish the office of municipal welfare officer (the *Almosenschaffner*). Following several years of dearth or famine, the Evangelical occupant of the office narrowed eligibility conditions for who would receive assistance based on ideas about the 'deserving' and 'undeserving' poor (although he did not use those terms).

Strasbourg's Evangelicals also worked with the *Rat* to reorganize the city's education system, among which was the creation of a new institution of higher education, the *Schola Argentoratensis*. Several of the city's most prominent theologians lectured to boys and young men who were poised to become social leaders throughout the Holy Roman Empire. Capito and Calvin's views that the natural environment was created by God for humanity and that natural events, both catastrophic and beneficial, are divine responses to human behaviour were inculcated in students who would become influential decision-makers over a wide area of the continent.

It is noteworthy that, as mentioned above, consequences of weather-related famine or dearth after 1525 do not seem to have included peasant plans for another rebellion. The high number of deaths among them and the measures taken by the powerful elite after the 1525 Peasants' War successfully suppressed any possibility of rebellion; at least, there is no evidence of further peasant organization against the social hierarchy. Perhaps in an attempt to gain the sympathies of Strasbourg's gardeners and to enlist the peasant community in Evangelical views, Capito articulated the view that God's pleasure could be found in an enthusiastic exploitation of nature for survival and economic benefit. This, according to him, was the appropriate human relationship to the natural environment. Wolfgang Capito and Jean Calvin, then – Capito regularly in *Hexemeron Dei opus* and Calvin embryonically in the 1539 edition of *Institutio Christianae Religionis* – provide evidence in support of Lynn White, Jr.'s thesis that Western Christianity is the intellectual and religious foundation of the current environmental crisis. Indeed, there is very little to be seen among the early sixteenth-century Evangelical community of Strasbourg that would contradict this confirmation.

Notes

1 The Reform in Strasbourg has been the subject of much outstanding historical research; see, among others, Lorna Jane Aubray, *The People's Reformation: Magistrates, Clergy, and Commons in Strasbourg, 1500–1598* (Oxford: Basil Blackwell, 1985); David Bagchi, 'Germany and the Lutheran Reformation,' in *The European Reformations*, ed. Alec Ryrie, Palgrave Advances (Basingstoke: Palgrave MacMillan, 2006), 13–35; Thomas A. Brady, Jr., *Ruling Class, Regime and Reformation at Strasbourg, 1520–1555* (Leiden: Brill, 1978); Thomas A. Brady, Jr., *German Histories in the Age of Reformation: 1400–1650* (Cambridge: Cambridge University Press, 2009); Thomas A. Brady, Jr., '"You Hate Us Priests": Anticlericalism, Communalism, and the Control of Women at Strasbourg in the Age of the Reformation,' in *Anticlericalism in Late Medieval and Early Modern Europe*, eds., Peter A. Dykema and Heiko A. Oberman (Leiden: Brill, 1993), 167–208; Miriam Usher Chrisman, *Strasbourg and the Reform: A Study in the Process of Change* (New Haven and London: Yale University Press, 1967); Miriam Usher Chrisman, 'Urban Poor in the Sixteenth Century: The Case of Strasbourg,' in *Social Groups and Religious Ideas in the Sixteenth Century*, Studies in Medieval Culture XIII (Kalamazoo: The Medieval Institute, Western Michigan University, 1978), 59–67; Miriam Usher Chrisman, *Lay Culture, Learned Culture: Books and Social Change in Strasbourg, 1480–1599* (New Haven and London: Yale University Press, 1982); Miriam Usher Chrisman, *Conflicting Visions of Reform: German Lay Propaganda Pamphlets, 1519–1530* (New Jersey: Humanities Press, 1996).

2 Capito's biographical details are from James M. Kittelson, *Wolfgang Capito: From Humanist to Reformer*, Studies in Medieval and Reformation Thought, XVII (Leiden: E.J. Brill, 1975) and Erika Rummel, 'Wolfgang Faber Capito Chronology,' *The Electronic Capito Project* <http://www.itergateway.org/capito/page2.html> [accessed 6 June 2023]. For an in-depth exploration of Calvin's time in Strasbourg, see Matthieu Arnold, ed., *John Calvin: The Strasbourg Years (1538–1541)*, trans. Felicity McNab (Eugene: WIPF & Stock, 2016).

3 Miriam Usher Chrisman, *Bibliography of Strasbourg Imprints, 1480–1599* (New Haven and London: Yale University Press, 1982), 285; Miriam Usher Chrisman, *Strasbourg and the Reform: A Study in the Process of Change* (New Haven and London: Yale University Press, 1967), 98, 99, 301. To distinguish the chapel inside the *Liebfrauenmünster* from the parish in the city (both named after Saint Lawrence), the chapel will be referred to in High Middle German (Sankt Laurentius) while the parish will be referred to in French (Saint Laurent). Chrisman, *Strasbourg and the Reform*, 98–9.

4 Esther Chung-Kim, *Economics of Faith: Reforming Poor Relief in Early Modern Europe* (OUP, 2021), 1–3.

5 Norman J. Wilson, 'Conceptions of Poor Relief in Sixteenth-Century Strasbourg,' *UCLA Historical Journal*, 8(2) (1987), 5–24.

6 Ibid.

7 Wilson, 11–12; Chung-Kim, 4; Katharina Schütz Zell, *Church Mother: The Writings of a Protestant Reformer in Sixteenth Century Germany*, trans. Elsie McKee (University of Chicago Press, 2007), 189, n. 22, quoted in Chung-Kim, 21, n.90; Chrisman, 'Urban Poor in the Sixteenth Century,' 59–67; much of Hackfurt's *Tagebuch* is available in Otto Winckelmann, *Das Fürsorgewesen der Stadt Strassburg*, Quellen und Forchunge zur Reformationsgeschichte (Leipzig, 1922), II, no. 40. See also Thomas A. Brady, Jr., *Protestant Politics: Jacob Sturm (1489–1553) and the German Reformation* (New Jersey: Humanities Press International, 1995), 117.

8 Chrisman, *Strasbourg and the Reform*, 115.
9 Aubray, 38–9.
10 For further discussion of the synod, see Kittelson, 195–6, among others.
11 Wolfgang Capito, *Institutiuncula in Hebream linguam. Autore Volphango Fabro Professor Theologiae* (Basel, 1516).
12 Survey for 1525 on the database www.Tambora.org, <www.tambora.org> [accessed 27 February 2023]; Daniel Specklin, *Les Collectanées de Daniel Specklin, chronique Strasbourgeoise du Seizième Siècle*, ed. Rodolphe Reuss, Fragments des Anciennes Chronique d'Alsace, II (Strasbourg: Librairie J. Noiriel, 1890), 511, Op. Cit. 2273 (Pp. P); P.F. Malachias Tschamser, ed., *Annales oder Jahrs-Geschicthen der Baarfüferen oder Minderen Brüdern S. Franc. ord. inßgemein Conventualen gevannt, zu Thann*, 2 vols (Colmar: Hoffmann, 1864), vol. II, 41; Hans Stolz, *Die Hans Stolz'sche Gebweiler Chronik. Zeugenbericht über den Bauernkreig am Oberrhein*, ed. Wolfram Stolz (Freiburg: Edition Stolz, 1979), 179–82.
13 Survey for 1526, 1527, and 1528 on the database www.Tambora.org, <www.tambora.org> [accessed 1 March 2023]; Jacques de Gottesheim, 'Les Éphémérides de Jacques de Gottesheim: Docteur en Droit, Prébendier du Grand-Choeur de la Cathédrale (1524–1543). Fragments publiés pour la première fois et annotés,' ed. Rodolphe Reuss, *Bulletin de la Société pour la Conservation des Monuments Historiques*, IIᵉ Serie, XIX, bk. I (Strasbourg: R. Schultz & Cie, 1898), 271; Stolz, 204–7; Tschamser, 44. See also Bibliothèque Municipale de Haguenau, MS 5.2, 213, cited in Claude Muller, *Chronique de la Viticulture Alsacienne au XVIe Siècle* (Riquewihr: Editions J.-D. Reber, 2005), 65; John D. Derksen, *From Radicals to Survivors: Strasbourg's Religious Nonconformists over Two Generations, 1525–1570* (Utrecht: Hes & De Graaf, 2002), 41; Xavier Mossmann, ed., *Chronique des dominicains de Guebwiller* (Guebwiller: G. Brückert; Colmar: L. Reiffinger; Strasbourg: Schmidt et Grucker, 1844), 145–6; Tschamser, II, 50–1; Mossmann, 169; Stolz, 224.
14 Survey for 1529 on the database www.Tambora.org, <www.tambora.org> [accessed 3 March 2023]; Specklin, p. 524, Op. Cit. 2306 (Exc. Sp); Stolz, 257–66; Tschamser, 56; Mossmann, 176–7, 180; Sébald Büheler, 'La Chronique Strasbourgeoise de Sébald Bühelers,' in *Fragments des Anciennes Chroniques d'Alsace*, ed. Léon Dacheux, I (Strasbourg: Schultz, 1887), 79n246; Jasper Schenk, '"Human Security" in the Renaissance? "Securitas", Infrastructure, Collective Goods and Natural Hazards in Tuscany and the Upper Rhine Valley/Human Security in der Renaissance? Securitas, Infrastruktur, Gemeinschaftsgüter und Naturgefahren in der Toskana und im Oberrheintal,' *Historical Social Research/Historische Sozialforschung*, Vol. 35, No. 4 (134), The Production of Human Security in Premodern and Contemporary History/Die Produktionvon Human Security in Vormoderne und Zeitgeschichte (2010), 223; Thomas A. Brady, Jr., *Protestant Politics: Jacob Sturm (1489–1553) and the German Reformation* (New Jersey: Humanities Press International, 1995), 80.
15 Survey for 1530 on the database www.Tambora.org, <www.tambora.org> [accessed 3 March 2023]; Mossmann, 183; Stolz, 272; Tschamser, II, 62–3; Bibliotheque Municipale de Haguenau, MS 5.2, 216, cited in Muller, 70.
16 Chrisman, 'Urban Poor in the Sixteenth Century: The Case of Strasbourg.'
17 Ibid.; Philip Kintner, 'Welfare, Reformation, and Dearth at Memmingen,' in *The Reformation of Charity: The Secular and the Religious in Early Modern Poor Relief*, ed. Thomas Max Safley, Studies in Central European Histories Series, XXX (Leiden: Brill, 2003), 74.
18 Survey for 1531 on the database www.Tambora.org, <www.tambora.org> [accessed 3 March 2023]; Mossmann, 186–7; Bibliothèque Municipale de Haguenau, MS 5.2, 216, cited in Muller, 74; Specklin, p. 529, Op. Cit. 2330 (Pp. Rh). Piton

mentions that according to Specklin's figures , 23,545 people received aid from the Ellendenherberg (the poorhouse), although that figure seems very high. Specklin, p. 529, Op. Cit. 2332 (Pp. P); Stolz, 281.

19 Survey for 1532 on the database www.Tambora.org, <www.tambora.org> [accessed 6 March 2023]; Tschamser, II, 68–70; Mossmann, 193–4.

20 Survey for 1533, 1534, and 1535 on the database www.Tambora.org, <www.tambora.org> [accessed 6 March 2023]; Gottesheim, 276; Specklin, p. 531, Op. Cit. 2341 (Pp. Sch); Tschamser, II, 75–7; Stolz, 331; Bibliotheque Municipale de Haguenau, MS 5.2, 218, cited in Muller, 80; Tschamser, II, 79; Stolz, 334; Mossmann, 202.

21 Survey for 1536 and 1537 on the database www.Tambora.org, <www.tambora.org> [accessed 8 March 2023]; Tschamser, II, 82; Mossman, 214; Friedrich Weissbass, *Weingewerbe und Weinleutenzunft im alten Basel* (Bâle, 1958), 291–5; Bibliothèque Municipale de Haguenau, MS 5.2, 220, cited in Muller, 84; Stolz, 343–5; Specklin, 532, Op. Cit. 2347 (Exc. Sp); Tschamser, II, 86; Mossmann, 215–6.

22 Survey for 1538 on the database www.Tambora.org, <www.tambora.org> [accessed 9 March 2023]; Stolz, 353; Franz-Josef, Mone, ed., *Quellensammlung der badischen Landesgeschichte*, 1 and 2 (Karlsruhe: C. Macklot, 1848) 2; Tschamser, II, 89, 91; Stolz, 358; Mossmann, 218–9.

23 Survey for 1539 on the database www.Tambora.org, <www.tambora.org> [accessed 9 March 2023]; Stolz, 362, 365, 369; Bibliothèque Municipale de Haguenau, MS 5.2, 222, cited in Muller, 92; F. Ritter, 'Incunables du grand séminaire de Strasbourg,' in *Archives de l'Église d'Alsace*, 1953–1954, 112, cited in Muller, 89; Johann Jacob Meyer, *La Chronique Strasbourgeoise*, ed. Rodolphe Reuss (Strasbourg: J. Noiriel, 1873),123–4; Specklin, 533, no. 2351; Jacob Trausch, 'Strassburgische Chronick,' in *Les Chroniques Strasbourgeoises de Jacques Trausch et de Jean Wencker*, ed. Léon Dacheux, Fragments des Anciennes Chroniques d'Alsace, III (Strasbourg: Imprimerie Strasbourgeoise and R. Schultz, 1892), 64; Mossman, 220–3; Tschamser, II, 92–3; E. Meininger, *Le Vieux Mulhouse*, vol. II, (Bader, 1897), 159, cited in Muller, 90; Weissbass, 292–3; Mossmann, 221.

24 Survey for 1540 on the database www.Tambora.org, <www.tambora.org> [accessed 10 March 2023]; Meyer, 124; Mone, 2, 143; Stolz, 373–80; Büheler, 84, no. 279; Specklin, 533–4, Op. Cit. 2353 (W); Mossmann, 224–8.

25 Oliver Wetter et al., 'The Year-Long Unprecedented European Heat and Drought of 1540 – A Worst Case,' *Climate Change*, 125 (2014), 357.

26 Wolfgang Capito, *Hexemeron Dei opus explicatum à Wolphgango Fa. Capitone Theologo. Cum Indice locupletißimo* (Argentorati: Vuendelinum Rihelium, September 1539). This study uses the volume at the Regensburg State Library, available digitally through the Bavarian State Library digital collection <http://reader.digitale-sammlungen.de/resolve/display/bsb11118009.html> [accessed 17 August 2023]. Capito baptized Rihel's eldest son Josias while the Rihel family were still in Hagenau and Rihel is known to have been educated in Latin. Chrisman, *Lay Culture, Learned Culture*, 14 and 25; CERL Thesaurus, 'Rihel, Wendelin (– 1555),' Consortium of European Research Libraries, last edited on 27 November 2006 <http://thesaurus.cerl.org/record/cni00025609> [accessed 17 August 2023].

27 Chrisman, *Lay Culture, Learned Culture*, 4, 6, 8, 18; Kittelson, Ch. 8; N. Scott Amos, *Bucer, Ephesians and Biblical Humanism: The Exegete as Theologian*, Studies in Early Modern Religious Tradition, Culture and Society, 7 (Cham: Springer International Publishing, 2015), 89; Arnold, 7; Marcel Fournier and Charles Engel, ed., *Les statuts et privilèges des universités françaises depuis leur fondation jusqu'en 1789, tome 4/1: Gymnase, Académie, Université de Strasbourg* (Paris: L. Larose, 1894), 50 and 53; Kittelson, 4.

28 John L. Thompson, ed., *Genesis 1–11*, Reformation Commentary on Scripture, I (Downer's Grove: InterVarsity Press, 2012), lxvi.
29 Capito, *Hexemeron*, fols. 17^v–18^r.
30 Capito, *Hexemeron*, fols. 100^v–1^r.
31 Sachiko Kusukawa, *The Transformation of Natural Philosophy: The Case of Philip Melanchthon* (Cambridge: Cambridge University Press, 1995), 33.
32 Ibid., 32–42.
33 Capito, *Hexemeron*, fols. 2345^v–5^r.
34 Capito, *Hexemeron*, fols. 293^{r-v}.
35 Capito, *Hexemeron*, fols. 283$^{r.-v}$.
36 Capito, *Hexemeron*, fol. 285^v.
37 Capito, *Hexemeron*, fol. 226^r.
38 Capito, *Hexemeron*, fols. 261^v–2^r and fols. 294^{r-v}.
39 Women are, according to Capito, equal to men but subject to them for the sake of order and from the sequence of creation; however, it must be noted that Capito did not extend similar authority to any creature created before man. Capito, *Hexemeron*, fols. 287^v–8^r.
40 Kusukawa, 59–61.
41 Ibid., 70.
42 Ibid., 154
43 Ibid., 151–4.
44 Ibid., 154–7.
45 Ibid., 158.
46 Ibid., 72, 149, and 187.
47 Capito, *Hexemeron*, fol. 101^r.
48 Biographical details for Jean Calvin are selected primarily from Bruce Gordon, *Calvin* (New Haven and London: Yale University Press, illustrated edition, 2011), as well as from Thomas I I.L. Parker, *John Calvin: A Biography* (Philadelphia: The Westminster Press, 1975) and Matthieu Arnold, ed., *John Calvin: The Strasbourg Years (1538–1541)*, trans. Felicity McNab (Eugene: WIPF & Stock, 2016).
49 Gordon, 115.
50 Richard A. Muller, *The Unaccommodated Calvin: Studies in the Foundation of a Theological Tradition* (Oxford: Oxford University Press, 1999), 120, 104, cited in Gordon, 91–2; Gordon, 92; Davis A. Young, *John Calvin and the Natural World* (Lanham: University Press of America, 2007), 59, 62, 108.
51 Jean Calvin, *Institutio Christianae Religionis nunc verè demum suo titulo respondens* (Argentorati per Vuendelinum Rihelium, Mense Augusto, Anno MDXXXIX), 6–7 and 29; available online at <https://www.digitale-sammlungen.de/en/view/bsb11118009?page=,1> (accessed 24 August 2023).
52 Calvin, 123.
53 Susan E. Schreiner, *The Theatre of His Glory: Nature & the Natural Order in the Thought of John Calvin* (Grand Rapids: Baker Academic, 1995), 4, 35–7.
54 Gordon, 108.
55 Brady, *Protestant Politics*, 116–25; Gerald Strauss, *Luther's House of Learning: Indoctrination of the Young in the German Reformation* (Baltimore and London: The Johns Hopkins University Press, 1978), 197; Boasting 485 years of humanist education on its website, the still-extant college is now known as *le gymnase Jean Sturm*. There are currently some 2000 pupils from kindergarten to the Baccalaureate, for which latter qualification students achieve a 100% success rate. <http://www.jsturm.fr/> [accessed 9 August 2023].
56 Brady, *Protestant Politics*, 119.

57 John William Adamson, *A Short History of Education* (Cambridge: Cambridge University Press, 2013; orig. 1919), 160–1; Marc Lienhard, 'Strasbourg in Calvin's Time,' in *John Calvin: The Strasbourg Years (1538–1541)*, ed., Mathieu Arnold, trans. Felicity McNab (Eugene: WIPF & STOCK, 2016), 7; Amos, 88–9; Wulfert Greef, *The Writings of John Calvin: An Introductory Guide* (Louisville: Westminster John Knox Press, 2008), 13.

References

Primary sources

Büheler, Sébald, 'La Chronique Strasbourgeoise de Sébald Bühelers,' in *Fragments des Anciennes Chroniques d'Alsace*, ed. Léon Dacheux, I (Strasbourg: Schultz, 1887).

Calvin, Jean, *Institutio Christianae Religionis nunc verè demum suo titulo respondens* (Argentorati per Vuendelinum Rihelium, Mense Augusto, Anno MDXXXIX); available online at <https://www.digitale-sammlungen.de/en/view/bsb11118009?page=,1> [accessed 24 August 2023].

Capito, Wolfgang, *Institutiuncula in Hebream linguam. Autore Volphango Fabro Professor Theologiae* (Basel: Froben, 1516).

——— *Hexemeron Dei opus explicatum à Wolphgango Fa. Capitone Theologo. Cum Indice locupletißimo* (Argentorati: Vuendelinum Rihelium, September 1539).

De Gottesheim, Jacques, 'Les Éphémérides de Jacques de Gottesheim: Docteur en Droit, Prébendier du Grand-Choeur de la Cathédrale (1524–1543). Fragments publiés pour la première fois et annotés,' in *Bulletin de la Société pour la Conservation des Monuments Historiques*, ed. Rodolphe Reuss, IIe Serie, XIX, bk. I (Strasbourg: R. Schultz & Cie, 1898).

Meyer, Johann Jacob, *La Chronique Strasbourgeoise*, ed. Rodolphe Reuss (Strasbourg: J. Noiriel, 1873).

Mone, Franz-Josef, ed., *Quellensammlung der badischen Landesgeschichte*, 1 and 2 (Karlsruhe: C. Macklot, 1848).

Mossmann, Xavier, ed., *Chronique des dominicains de Guebwiller* (Guebwiller: G. Brückert; Colmar: L. Reiffinger; Strasbourg: Schmidt et Grucker, 1844).

Specklin, Daniel, *Les Collectanées de Daniel Specklin, chronique Strasbourgeoise du Seizième Siècle*, ed. Rodolphe Reuss, Fragments des Anciennes Chronique d'Alsace, II (Strasbourg: Librairie J. Noiriel, 1890).

Stolz, Hans, *Die Hans Stolz'sche Gebweiler Chronik. Zeugenbericht über den Bauernkreig am Oberrhein*, ed. Wolfram Stolz (Freiburg: Edition Stolz, 1979).

Trausch, Jacob, 'Strassburgische Chronick,' in *Les Chroniques Strasbourgeoises de Jacques Trausch et de Jean Wencker*, ed. Léon Dacheux, Fragments des Anciennes Chroniques d'Alsace, III (Strasbourg: Imprimerie Strasbourgeoise and R. Schultz, 1892).

Tschamser, P.F. Malachias, ed., *Annales oder Jahrs-Geschicthen der Baarfüferen oder Minderen Brüdern S. Franc. ord. inßgemein Conventualen gevannt, zu Thann*, 2 vols (Colmar: Hoffmann, 1864).

Zell, Katharina Schütz, *Church Mother: The Writings of a Protestant Reformer in Sixteenth Century Germany*, ed. and trans. Elsie McKee (Chicago: University of Chicago Press, 2007).

Secondary sources

Adamson, John William, *A Short History of Education* (Cambridge: Cambridge University Press, 2013; orig. 1919).

Amos, N. Scott, *Bucer, Ephesians and Biblical Humanism: The Exegete as Theologian*, Studies in Early Modern Religious Tradition, Culture and Society, 7 (Cham: Springer International Publishing, 2015).

Arnold, Matthew, ed., *John Calvin: The Strasbourg Years (1538–1541)*, trans. Felicity McNab (Eugene: WIPF & Stock, 2016).

Aubray, Lorna Jane, *The People's Reformation: Magistrates, Clergy, and Commons in Strasbourg, 1500–1598* (Oxford: Basil Blackwell, 1985).

Bagchi, David, 'Germany and the Lutheran Reformation,' in *The European Reformations*, ed. Alec Ryrie, Palgrave Advances (Basingstoke: Palgrave MacMillan, 2006), 13–35.

Brady, Thomas A., Jr., *Ruling Class, Regime and Reformation at Strasbourg, 1520–1555* (Leiden: Brill, 1978).

——— '"You Hate Us Priests": Anticlericalism, Communalism, and the Control of Women at Strasbourg in the Age of the Reformation,' in *Anticlericalism in Late Medieval and Early Modern Europe*, eds., Peter A. Dykema and Heiko A. Oberman (Leiden: Brill, 1993), 167–208.

——— *Protestant Politics: Jacob Sturm (1489–1553) and the German Reformation* (New Jersey: Humanities Press International, 1995).

——— *German Histories in the Age of Reformation: 1400–1650* (Cambridge: Cambridge University Press, 2009).

Chrisman, Miriam Usher, *Strasbourg and the Reform: A Study in the Process of Change* (New Haven and London: Yale University Press, 1967).

——— 'Urban Poor in the Sixteenth Century: The Case of Strasbourg,' in *Social Groups and Religious Ideas in The Sixteenth Century*, Studies in Medieval Culture XIII (Kalamazoo: The Medieval Institute, Western Michigan University, 1978), 59–67.

——— *Bibliography of Strasbourg Imprints, 1480–1599* (New Haven and London: Yale University Press, 1982).

——— *Lay Culture, Learned Culture: Books and Social Change in Strasbourg, 1480–1599* (New Haven and London: Yale University Press, 1982).

——— *Conflicting Visions of Reform: German Lay Propaganda Pamphlets, 1519–1530* (New Jersey: Humanities Press, 1996).

Chung-Kim, Esther, *Economics of Faith: Reforming Poor Relief in Early Modern Europe* (Oxford: Oxford University Press, 2021).

Derksen, John D., *From Radicals to Survivors: Strasbourg's Religious Nonconformists over Two Generations, 1525-1570*, Bibliotheca Humanistica & Reformatorica, 61 (Utrecht: Hes and de Graaf, 2002).

Fournier, Marcel, and Charles Engel, ed., *Les statuts et privilèges des universités françaises depuis leur fondation jusqu'en 1789, tome 4/1: Gymnase, Académie, Université de Strasbourg* (Paris: L. Larose, 1894).

Greef, Wulfert, *The Writings of John Calvin: An Introductory Guide* (Louisville: Westminster John Knox Press, 2008).

Gordon, Bruce, *Calvin* (New Haven and London: Yale University Press, illustrated edition, 2011).

Kintner, Philip, 'Welfare, Reformation, and Dearth at Memmingen,' in *The Reformation of Charity: The Secular and the Religious in Early Modern Poor Relief*, ed. Thomas Max Safley, Studies in Central European Histories Series, XXX (Leiden: Brill, 2003), *Generations, 1525–1570* (Utrecht: HES & DE GRAAF Publishers BV, 2002).

Kittelson, James M., *Wolfgang Capito: From Humanist to Reformer*, Studies in Medieval and Reformation Thought, XVII (Leiden: E.J. Brill, 1975).

Kusukawa, Sachiko, *The Transformation of Natural Philosophy: The Case of Philip Melanchthon* (Cambridge: Cambridge University Press, 1995).

Lienhard, Marc, 'Strasbourg in Calvin's Time,' in *John Calvin: The Strasbourg Years (1538–1541*, ed., Mathieu Arnold, trans. Felicity McNab (Eugene: WIPF & STOCK, 2016).

Muller, Claude, *Chronique de la Viticulture Alsacienne au XVIe Siècle* (Riquewihr: Éeditions J.-D. Reber, 2005).

Parker, Thomas H.L., *John Calvin: A Biography* (Philadelphia: The Westminster Press, 1975).

Rummel, Erika, 'Wolfgang Faber Capito Chronology,' *The Electronic Capito Project* <http://www.itergateway.org/capito/page2.html> [accessed 6 June 2023].

Schenk, Jasper, '"Human Security" in the Renaissance? "Securitas," Infrastructure, Collective Goods and Natural Hazards in Tuscany and the Upper Rhine Valley/Human Security in der Renaissance? Securitas, Infrastruktur, Gemeinschaftsgüter und Naturgefahren in der Toskana und im Oberrheintal,' *Historical Social Research/Historische Sozialforschung*, Vol. 35, No. 4 (134), The Production of Human Security in Premodern and Contemporary History/Die Produktionvon Human Security in Vormoderne und Zeitgeschichte (2010).

Schreiner, Susan E., *The Theatre of His Glory: Nature & the Natural Order in the Thought of John Calvin* (Grand Rapids: Baker Academic, 1995).

Strauss, Gerald, *Luther's House of Learning: Indoctrination of the Young in the German Reformation* (Baltimore and London: The Johns Hopkins University Press, 1978).

Thompson, John L., ed., *Genesis 1–11*, Reformation Commentary on Scripture, I (Downer's Grove: InterVarsity Press, 2012).

Weissbass, Friedrich, *Weingewerbe und Weinleutenzunft im alten Basel* (Basel: Zunft zu Weinleuten, 1958).

Wetter, Oliver, Christian Pfister, Johannes P. Werner, Eduardo Zorita, Sebastian Wagner, Sonia I. Seneviratne, Jürgen Herget, Uwe Grünewald, Jürg Luterbacher, Maria-Joao Alcoforado, Mariano Barriendos, Ursula Bieber, Rudolf Brázdil, Karl H. Burmeister, Chantal Camenisch, Antonio Contino, Petr Dobrovolný, Rüdiger Glaser, Iso Himmelsbach, Andrea Kiss, Oldřich Kotyza, Thomas Labbé, Danuta Limanówka, Laurent Litzenburger, Øyvind Nordl, Kathleen Pribyl, Dag Retsö, Dirk Riemann, Christian Rohr, Werner Siegfried, Johan Söderberg, and Jean-Laurent Spring, 'The Year-Long Unprecedented European Heat and Drought of 1540 – A Worst Case,' in *Climate Change*, 125 (2014). doi: 10.1007/s10584-014-1184-2.

Wilson, Norman J., 'Conceptions of Poor Relief in Sixteenth-Century Strasbourg,' *UCLA Historical Journal*, 8(2) 1987, 5–24.

Winckelmann, Otto, *Das Fürsorgewesen der Stadt Strassburg, vor und nach der Reformation bis zum Ausgang des sechzenhten Jahrhunderts*, Quellen und Forchunge zur Reformationsgeschichte (Leipzig: Verein für Reformationgeschichte, 1922).

Young, Davis A., *John Calvin and the Natural World* (Lanham: University Press of America, 2007).

Digital sources

CERL Thesaurus, 'Rihel, Wendelin (– 1555),' Consortium of European Research Libraries, last edited on 27 November 2006 <http://thesaurus.cerl.org/record/cni00025609> [accessed 17 August 2023].

Conclusions

With this research, I have demonstrated the strength of two hypotheses: first, that the unreliable weather conditions characteristic of the Spörer Minimum exerted such severe economic stress on Alsatians that the bad weather contributed to their enthusiastic acceptance of theological developments we now refer to as the Protestant Reformation. Therefore, far from being a neutral backdrop, changes in the natural environment – specifically, in the climate – must be listed among contributing factors for the success of the Protestant Reformation. While my evidence is focused on Alsace, unstable and unreliable weather conditions were common across southern Germany and, indeed, throughout northern Europe, which provides solid grounds to generalize from the bioregion to the wider population. My second hypothesis focuses more closely on the new Christian representation of nature which emerged as part of the Reformed theologies. Specifically, I posit that the weather-related social turmoil of the period was the fertile ground from which Evangelical views emerged which explicitly declared the world to be devoid of any intrinsic value to God, based on the view that God created the earth exclusively in order to meet human needs until salvation is achieved.

To support my arguments, I have described almost seven decades of weather conditions in Alsace, described some social responses to those conditions, and investigated three different denotations the natural world had for early sixteenth-century people in Strasbourg and Alsace. In three case studies, I have explored *Die Emeis*, a collection of sermons by Johannes Geiler von Kaysersberg; *Ein Fast schon büchlin* from Clemens Zyegler; the peasant army's *XII Artikels;* and *Hexemeron Dei opus* by Wolfgang Capito, to identify and explore similarities and differences in their religious representation of the natural environment and the appropriate human relationship to it. These conclusions summarize my weather report, review some of the social consequences of that weather, and offer an evaluation of changes in the religious representations of nature from Geiler through Zyegler to Capito.

DOI: 10.4324/9781003262411-6

Overview of the weather from 1473 to 1540 and some social consequences of that weather

The weather of the Upper Rhine Valley from 1473 to 1540 was challenging for the humans and animals living there because the unreliable and uncertain weather conditions characteristic of the Spörer Minimum reached their zenith during this period. The weather report for those 68 years is revealing: during the period covered, harvests were reported to be above in average 18 years (1479, 1482, 1483, 1484, 1495, 1499, 1505, 1506, 1510, 1514, 1523, 1526, 1527, 1528, 1532, 1536, 1537, and 1539). They were noted as average or missing from the archive entirely (which assumes an average harvest, neither abundant nor poor; not worthy of mention either way) in 21 years (1473, 1474, 1475, 1478, 1485, 1493, 1496, 1497, 1498, 1502, 1504, 1508, 1513, 1518, 1519, 1521, 1522, 1525, 1530, 1533, and 1538). The harvests were described as poor in 17 years (1476, 1477, 1480, 1481, 1486, 1487, 1488, 1489, 1490, 1491, 1492, 1507, 1509, 1524, 1529, 1534, and 1540) and as having failed outright in 12 years (1494, 1500, 1501, 1503, 1511, 1512, 1515, 1516, 1517, 1520, 1531, and 1535).

Since Alsatian harvests failed in 12 of 68 years (18%) and were poor for 17 years (25%), inadequate harvests were the agricultural results for 29 of 68 years (43%). However, averaging the failed and poor harvests over 68 years hides periods of heightened stress on Alsatians from the natural world which a closer scrutiny reveals. Of the 36 harvests from 1473 to 1508, four failed outright (11%). However, of the 16 harvests from 1509 to 1524, six failed completely (37%), while from 1525 to 1540, the failure rate dropped to 2 out of 16 harvests (13%).

This ebb and flow of weather conditions unpredictably exacerbated rigid economic demands on the peasants by social superiors, creating unforeseeable and erratic conditions of severe stress for Alsatian peasants and the urban poor of Strasbourg. The correlation between weather-related harvest failure and the *Bundschuh* rebellions demonstrates that weather conditions contributed to social instability: consistently after two or more years of weather-related harvest disappointments (whether poor or failed), peasants would plan and organize an uprising. That is, following poor harvests for six consecutive years (from 1486 to 1492), including three years of dearth (scarce food supply) in 1490, 1491, and 1492, plans for the first *Bundschuh* rebellion were organized in spring 1493. The second *Bundschuh* rebellion was planned for spring 1502, following two back-to-back years of harvest failure (1500 and 1501, with regional famine (drastic food shortage) in 1501). The third rebellion was planned for spring 1513 after another two consecutive years of harvest failure in 1511 and 1512, when the harvest failure of 1512 led to dearth in the bioregion. Plans for the fourth *Bundschuh* developed for autumn 1517, following harvest failures and famine in 1515, 1516, and 1517. There was partial deviation from the pattern when, after only a single harvest failure and famine in 1524, Alsatian peasants contributed plans, resources, and lives to the great 1525 German peasant uprising. However, in this instance, joining a widespread rebellion against an exploitative and oppressive social hierarchy may have been more appealing than waiting for further stressors to arrive from the natural environment.

Weather conditions were not the only factor involved in this social turbulence; institutionalized and deepening social inequality also played its part in insuring that the economic repercussions of harvest failure were felt most keenly by the poor. Nevertheless, social inequality on its own did not trigger rebellion among the poor, as demonstrated by the years of good weather in the region during which social inequality was unrelieved. Plans or outbreaks of armed rebellion followed on the heels of poor weather conditions, as their economic consequences imposed unbearable stress on the more vulnerable members of society.

Luther opens a door

Following three years of famine (1515, 1516, and 1517), repression of the last *Bundschuh* rebellion in autumn 1517, continuing threat of weather-related harvest failure and Strasbourg's dancing plague in summer 1518, news of Martin Luther's theological innovations reached Strasbourg and Alsace. With this new development from the cultural sphere, social programmes which made sense according to a Roman Catholic worldview (for example, weather rogations or the exorcism of a storm) lost cultural authority. By 1524, individuals refused to participate in rogations from Strasbourg's *Liebfrauenmünster*, and the *Rat*'s ban of the mass within city walls in 1529 solidified the city's institutional rejection of Roman Catholic interpretative power.

Luther's religious insights opened the door to several competing yet culturally congruent explanations for, among other things, the repeated hardships and disasters created by the unstable weather conditions characteristic of the Spörer Minimum. Those who were convinced by Luther's challenge to Catholic cultural authority, whether they accepted radical, Reformed, or other religious views in its place, also rejected the Roman Catholic view of the natural environment (along with impious religious officials, institutional corruption, and other shortcomings). Participation dropped in Catholic apotropaic rituals intended to create supportive weather conditions and, when bad weather caused another poor harvest and famine in 1524, many who had not engaged with the *Bundschuh* rebellions were inspired by the new exposition of the Bible to join the German Peasants' War of 1524 and 1525. Their faith in this new exegesis, exemplified by the preaching and pamphlets of Clemens Zyegler, gave them confidence in attempting to establish by force of arms an egalitarian agricultural community which, they believed, was God's will for humanity on earth. They were not interested in learning moral lessons from the denizens of nature (as preached by Geiler), so much as in turning wilderlands into farms.

Harvests failed again due to foul weather conditions in 1529, 1531, 1534, 1535, and 1540, leading to famine in 1531 and 1535. After defeat in 1525, however, social conditions had changed: the human and financial resources needed for peasant rebellion were no longer available and peasant culture was under assault. In the rural regions of Alsace, Roman Catholic cultural authority was re-imposed by imperial legislation requiring peasants to attend Mass on a regular basis, regardless of their personal beliefs, and radical religious views were outlawed. In Strasbourg, prominent radical theologians like Zyegler were silenced and excluded from responsible positions in the community. Provoked by repeated *Bundschuh* rebellions and the 1525 Peasants' War, Imperial authority developed a stronger presence in the region by imposing a rationalized system of fines for the benefit of the crown, a system which is perceived today as the forerunner of modern national taxation systems. Rebellions which were provoked – in part – by repeated and long-term material insecurity caused by destructive weather events characteristic of the Spörer Minimum thereby indirectly contributed to the birth and development of the nation-state, a long-term process whose beginnings were just underway in the early sixteenth century.

Comparing views of nature

The religious views of Evangelical Reformers became politically dominant in the Free Imperial City of Strasbourg after the military defeat of the peasants and their radical views in 1525; many scholars have explored this process (see Chapter 4, note 1). Evangelical political ascendancy, however, did not give Reform leadership the cultural influence they desired; contentious views continued to find public expression until 1533 and private circulation thereafter. For Reform leaders, establishing themselves as culturally preeminent meant appealing to non-Evangelicals, including those who held radical religious views; while those who remained stalwart Catholics were unlikely to question their received knowledge, the radically religious had demonstrated themselves open to alternative interpretations of the Bible. One possible way of appealing to peasants who accepted views like those of Zyegler, where God had created the earth to be the material foundation of an agricultural Kingdom of Heaven, might be to expand upon that view by claiming that God gave humans the whole world solely in order to meet our material needs, and that there is no value to be found in nature beyond its exploitation. Although we cannot confirm his motivations in doing so, Capito articulated this view in *Hexemeron Dei opus* with his exegesis of Genesis, where he declared that the entire earth – all the natural world – was created by God exclusively for humans to use until salvation is achieved. With the establishment of the *Schola Argentoratensis* in 1538, Capito's perspective on the natural environment was taught to the sons of Europe's elite.

There are both similarities and differences among these three Christian representations of nature, reflected in findings of both common ground and dramatic differences among the Catholic, radical, and Evangelical authors studied. My research necessarily explores the writings of specific individuals; it is important to remember, though, that these particular individuals were leaders of their communities, lending plausibility to the supposition that the religious views they expressed were widely shared. It is equally important to remember, when considering the prevalence of each distinct view in Strasbourg and Alsatian culture, that while they have been presented here discretely and in chronological order, this is an artificial separation: such divergent representations of and approaches to nature would have been contested among people sharing the same time and space. The views of the radicals were developing in rural parts of Alsace while Catholic preachers filled the pulpits of local churches; during the heavy drought of 1540, a year after Capito published *Hexemeron Dei opus*, Catholics in Guebwiller gave a mass for rain, as well as a separate mass for a good harvest of wine. Although each group sought to impose its vision of a Godly society throughout Alsace, none were able to achieve the full control of the social order they desired during this period of study. Consensus about the righteous social order was not realized.[1]

All four text sources represent the natural environment as organized by a common principle: divine order. There are other similarities among the authors included in this book, such as their common use of Scripture as the basis of their exhortations and their fervent desire to convince their audience of the righteousness of their perspective. Such commonalities, though, are a feature of the selection criteria established at the beginning of this project, particularly the period and location. Order is a basic principle found in all human cultures, as it is a necessary element in the development of a coherent worldview. It is not unexpected to find that these three men represented the world as an ordered place; of greater interest are the differences in the type of order each offered to his audience.

According to Johannes Geiler von Kaysersberg, God's natural order was established through the design of each creature, which was then given life by God. To the preacher, the presence of a tiny share of God's wisdom in each creature gave them value in God's eyes and meant that humans could receive moral teachings from animals, whose lack of free will meant they unselfconsciously behaved in accordance with God's will. Although Geiler asserted that God's ultimate authority meant all events occurred with His permission and for the ultimate good of humanity, the Roman Catholic preacher was also prepared to recognize that some natural events, such as the attack of a wolf, could be due either to God's lupine design and material causes, such as starvation or illness, or to supernatural intervention (whether angelic or diabolic). While Geiler acknowledged that supernatural entities might disrupt the integrity of God's design for benign or nefarious purposes, he also suggested that, if demons troubled humans, superior authorities, both secular and/or spiritual, could be called upon to restore order.

Geiler's representation of the natural world bears an unmistakable similarity to the feudal social order, where an appeal to a superior power against the transgressions of an intermediate power would have been instantly recognizable by his audience as a traditional means of defence against external danger. In that culture, this was the predominant way of (re)establishing order. Affirming a divinely ordered hierarchy to be the structure of the natural world may have been a strategy by Geiler to suggest that the social hierarchy was equally divinely ordered, although it is difficult to confirm such speculation. In his sermons for Lent 1509, the preacher advised his audience to 'be like the ant,' although the pastoral context implies he meant the advice in a moral sense rather than as an exhortation to acquiesce to social subordination. However, the conceptual distance between being humble before one's spiritual Lord in heaven and being humble before one's temporal Lord on earth is small. While Geiler did not bluntly state 'Oh, you peasants, be as the humble ant and remain uncomplaining in your lowly position at the bottom of the social order!,' it is noteworthy that he strongly asserted a hierarchical natural order at a time when Alsace had experienced at least two plans for armed rebellion (the *Bundschuh* rebellions of 1493 and 1502) during his tenure as Strasbourg's cathedral preacher.

Order was equally important to Clemens Zyegler in his *Ein fast schon büchlin in welche(n) yederman findet ein hellen und claren verstandt von dem leib und blůt Christi*, and to the peasant authors of the *XII Artikels*, but the structure of the order they valued was significantly different from Geiler's. In his pamphlet discussing the Eucharist, Zyegler recognizes God as exercising ultimate authority over Creation, but suggests that God's order is to be found in an egalitarian agricultural community. In this ideal community, individuals might guide each other towards righteousness, as one would guide domestic livestock back to its owner, but providing such guidance would not entitle the guide to a commensurate rise in social status. Zyegler's egalitarianism is limited to humanity; there is no suggestion that animals, for example, have any claims to rights. Indeed, Zyegler's representation of nature is focused so exclusively on agriculture and its accompanying activities that it is difficult to assess his views of the environment beyond the farm. However, in light of Zyegler's statement that a man is better than a cow, it might be safe to assume that he accepts a hierarchical order of the natural world as God's intention, with an egalitarian human society tucked peaceably within it.

That such an understanding of a Godly order was shared widely by peasant communities throughout southwestern Germany is implied by the *XII Artikels* written by the Swabian rebel representatives and distributed throughout the area of the 1525 rebellion, with variations repeatedly presented to authorities (including those of Alsace). Once again, final authority is ascribed to God; the authors portray God's preferred social order as one found in an interdependent and neighbourly agricultural community. They also call upon temporal Lords to acquiesce in restoring Godly order by allowing peasants to hunt wild animals which trespass onto croplands. In this view of the environment, wilderness and wild animals have a place, but it is outside the order that God intended for humanity. This perspective is agro-centric, and hints towards a view of wilderness as intrinsically without value, although peasants were reputedly quick to ignore their farms and turn to the wilderness for their material needs when necessary. The question of peasant views of the natural environment would benefit from further research with a wider geographical and temporal scope.

It is not as challenging to discover Wolfgang Capito's appreciation of a hierarchical natural order in Creation, thanks to the many times in *Hexemeron Dei opus* he repeats his view that God gave humanity full authority over the earth and its disposal. In his 1539 exegesis of Genesis, Capito sharply truncates the extensive medieval hierarchy of both material and spiritual beings as culturally transmitted by the concept of the Great Chain of Being; many of the supernatural beings Geiler had positioned as mediators between humanity and God disappear from the Evangelical's view of Creation. The devil and his demons, as well as most saints and many angels, are removed from explanations of natural events, processes, and forces; the communal rituals performed as appeals to such mediators were prohibited in Evangelical Strasbourg. Capito's Reforming views do not allow for any intermediaries between the human heart and God, including with respect to the natural realm.

Capito's perception of the righteous natural order also awards humans full entitlement to consume the earth's bounty without regard for spiritual consequences. That is, any intrinsic value which might have been ascribed to the natural environment is absent from his emphatic exposition of human salvation as the single purpose for which the earth was created. Unlike Geiler, Capito does not find moral teachings among the animals for humans to consider, nor does he allocate particles of God's wisdom to the beasts of the earth. Instead, in his expansions upon humanity's divinely ordained dominion over the earth, Capito includes an articulate justification for a ceaseless exploitation of the natural world: as long as there are human beings on the earth, God will render the earth available to meet humanity's needs. Capito's Reformed view of God's order on earth is very convenient for human beings; perhaps, with this view, Capito sought to attract radical Christian believers to Evangelical theology by integrating an approach to the natural world which he imagined likely to be received with favour by peasants.

Along with declaring the earth to be at humanity's entire disposal, however, Capito also asserts that God's omnipotent and omnipresent will imbues and motivates every aspect of Creation except humanity. Both Capito and Jean Calvin declare that God, through His will, untiringly maintains the existence of the world and, therefore, is omnipresent in the natural environment. Since He is responsible for every beneficial and destructive event, those natural events which affect human individuals and communities are to be understood as rewards and punishments from God. Just as there is no intrinsic value to the natural environment in Capito's exegesis, nothing occurs which is outside of God's immediate and intentional design. There is only God, working in ineffable ways. While Geiler can attribute the attack of a wolf to starvation, Capito would only be able to attribute such an attack to

God's will (although Capito did not discuss such phenomena as wolf attacks in *Hexemeron Dei opus*). Professing God's immediate presence in the natural world while asserting the natural world was created for human consumption draws a striking parallel between the natural world and the Eucharist, within which, when consecrated, Christ's special presence is recognized by the faithful. However, Capito did not discuss the theological implications of this similarity, limiting himself to presenting God as simultaneously closer to humanity because of his attentive immanence in nature and as having given the natural world to human beings.

Climate and religion

While the direct impact of weather on social events from 1493 to 1540 is comprehensible, the interactive, hybrid model of socio-economic metabolism reminds us that culture is an autonomous sphere of causation and activity. Developments in human culture may arise spontaneously, in response to human experience through their biophysical structures, and/or in response to developments in the natural sphere of causation which have an effect on human biophysical structures. While activity and responses from within the cultural sphere of recursive communication may arise independently of the natural environment (and often do), the cultural sphere interacts with the natural sphere through humanity's biophysical structures. This makes the possibility of a relationship between the two spheres analytically available.

One place where such a relationship may be presumed is in the eagerness with which many Alsatians greeted Luther's religious insights. His reinterpretation of the primary Christian text served to offer a culturally valid alternative to the status quo, in that (initially) he confronted the social structures which had developed in Catholic Europe without betraying the root principles upon which they stood. The cultural straitjacket by which Alsatian peasants and the urban poor of Strasbourg were bound excluded alternative meanings for natural events and led to profound psychological consequences such as the dancing plague when the weather did not support adequate food supply. Until Luther, the only way to understand or respond to poor weather leading to harvest failure and famine was controlled by Catholic religious representatives, among whom the topmost – such as the Bishop of Strasbourg – wielded significant financial and political power within a hierarchical social system. When confronted by a diminished ability to access the material resources necessary to meet financial demands imposed by social superiors, peasants first engaged with the ritual means available to them, then attempted to change the social hierarchy and, when that failed, greeted Luther's cultural innovation with enthusiasm.

The nature and direction of Luther's theological revelations could not have been anticipated. However, once their conceptual foundations had been understood by others, the opportunities they held were rapidly exploited by those who stood most to benefit from changes in the status quo. Was this

willingness by peasants and their allies to abandon the spiritual and social
security of Catholic orthodoxy in favour of reform theologies partially a
consequence of repeated experiences of weather-related economic hardship?
Evidence about the timing of rebellions, summarized earlier, shows that the
weather was a contributing factor.

Another place where the unsettled weather of late fifteenth- and early
sixteenth-century Alsace may have influenced changes in the cultural representation of nature may be seen in Capito's great appreciation of the role of
natural resource exploitation for human welfare. In Alsatian society of the
early sixteenth century, battered by the unreliable weather conditions typical
of the Spörer Minimum, religious views which expressed explicit admiration
for human exploitation of the natural world may have been warmly received.
The relatively brief time frame of my research does not permit an exploration
of the spread of Capito's views and the consequences, if any, for the work
undertaken by those who may have entertained them (for example, changes
in degree or type of natural resource exploitation).

Along with changes in the religious representation of the natural world
provoked by stress from the natural sphere of causation, theological developments in Strasbourg affected culturally meaningful programmes intended to
control the environment, particularly the ritual attempts to direct it. Rogations, prophylactic bell-ringing, blessings of destructive weather, and other
similar rites vanished from Strasbourg (although they accompanied the re-
imposition of the Roman rites in rural Alsace), to be replaced with Evangeli-
cally approved methods of approaching God, such as Capito's suggestions of
fasting or prayer.

Weber and White

At the outset, I engaged with the arguments of Max Weber and Lynn White,
Jr., and their hypotheses demand a clear response. That is, I have demonstrated that the hypothesis of the first is challenged by my research, while
that of the second finds support. Max Weber's view of Protestantism as a
rational religion whose explosion onto the European stage irretrievably shat-
tered what he saw as the irrational, magic-riddled miasma of late medieval
Catholicism is not upheld here. Geiler's sermons show him exercising a weak
form of rationality, in that he demonstrates a fully rational approach to the
natural world, despite holding irrational beliefs about it. That is to say, Geiler
is reasoned in his approach to a natural world which, for him, includes the
existence of spiritual beings along with birds, trees, and lions. Founded upon
classically based natural philosophy, Catholic doctrine, and the application
of logic to theological issues, his representation of the natural environment
is consistent and incorporates a wide range of possible causes for natural
events, forces, and processes, some of which are spiritual and some material.

Clemens Zyegler shows a comparable rationality, although his focus
in *Ein fast schon büchlin* is narrowly on the farm. Within that focus, he

demonstrates knowledge of farm operations that follow a rational process, and he reasons logically from them to demonstrate spiritual truths. Zyegler, therefore, also shows a weak form of rationality while expressing his radical religious views.

Rather than demonstrating strong rationality, as suggested by Weber, Wolfgang Capito also demonstrates a weak form of rationality. That is, his representation of the natural world is also consistently rational in the context of irrational beliefs held about it. Capito's views are extrapolated in a rational fashion from his understanding of Scripture, according to which God's omniscience and omnipotence support His will as the primary cause of all events, forces, and processes in the environment. It could even be suggested that Capito's rationality was a little weaker than Geiler's or Zyegler's, due to his recommendation that humanity avoid intellectual interest in or speculation about the natural environment and his insistence that natural events, forces, and processes were perforce a moral lesson from God. With advice like this, Capito discouraged the development of a rational understanding of the natural environment.

Lynn White, Jr.'s indictment of Western Christianity as instrumental in the current environmental crisis finds strong support in my research, particularly in Zyegler's *Ein fast schon büchlin* and Capito's *Hexemeron Dei opus*. Zyegler's valorization of the agricultural community as beloved of God implies a disdain for uncultivated wilderness and a lack of appreciation for its role in a wide number of ecologies. Capito's repeated assertions that God bestowed the natural world upon humanity for our use are incriminating articulations of an approach to the environment whose limitations are within sight today. Questions arise, though, about Capito's motivations for this approach, and the juxtaposition of this bestowal with a claim (shared by Calvin) that God is omnipresent in the natural world creates some confusion.

Concepts and ideas do not develop in isolation from the people who create them and who often use them to justify their decisions and behaviour. This exploration of changes in the concept of *nature* in early sixteenth-century Strasbourg and Alsace has served to illuminate some foundations of a powerful idea which, while it has supported Western society's unprecedented expansion, is also reaching its visible limits. Namely, the notion that humans are entitled to exploit the natural environment to exhaustion. It is increasingly obvious that this view is no longer of service to human survival; as a culture, we must change our minds about nature.

Note

1 Xavier Mossmann, ed., *Chronique des dominicains de Guebwiller* (Guebwiller: G. Brückert; Colmar: L. Reiffinger; Strasbourg: Schmidt et Grucker, 1844), 225.

References

Primary sources

Capito, Wolfgang, *Hexemeron Dei opus explicatum* à *Wolphgango Fa. Capitone Theologo* (Strasbourg: Vuendelinum Rihelium, 1539).
Geiler von Kaysersberg, Johannes, *Die Emeis*, ed. Johannes Pauli (Strasbourg: Johann Grüninger, 1517).
Mossmann, Xavier, ed., *Chronique des dominicains de Guebwiller* (Guebwiller: G. Brückert; Colmar: L. Reiffinger; Strasbourg: Schmidt et Grucker, 1844).
Zyegler, Clemens, *Ein Fast schon büchlin in welchen yederman findet ein hel-len und claren verstandt von dem leib und blût Christi* (Strasbourg: Johanem Schwan(?), 1525).

Secondary sources

Weber, Max, 'Science as a Vocation', in *From Max Weber: Essays in Sociology*, trans. and ed. Hans Heinrich Gerth and C. Wright Mills (New York: Oxford University Press, 1946), pp. 129–156; first pub. as 'Wissenschaft als Beruf', in *Gesammlte Aufsaetze zur Wissenschaftslehre* (Tübingen 1922), 524–55.
White, Lynn, Jr., 'The Historical Roots of Our Ecological Crisis,' *Science*, New Series, 155:3767 (10 March 1967), 1203–7.

Index

Note: Page numbers followed by "n" denote endnotes.

For Product Safety Concerns and Information please contact our
EU representative GPSR@taylorandfrancis.com Taylor & Francis
Verlag GmbH, Kaufingerstraße 24, 80331 München, Germany